Physical Chemistry: A Very Short Introduction

VERY SHORT INTRODUCTIONS are for anyone wanting a stimulating and accessible way into a new subject. They are written by experts, and have been translated into more than 40 different languages.

The Series began in 1995, and now covers a wide variety of topics in every discipline. The VSI library now contains over 350 volumes—a Very Short Introduction to everything from Psychology and Philosophy of Science to American History and Relativity—and continues to grow in every subject area.

Very Short Introductions available now:

THE VIKINGS Julian Richards
VIRUSES Dorothy H. Crawford
WITCHCRAFT Malcolm Gaskill
WITTGENSTEIN A. C. Grayling
WORK Stephen Fineman

WORLD MUSIC Philip Bohlman
THE WORLD TRADE
    ORGANIZATION Amrita Narlikar
WRITING AND SCRIPT
    Andrew Robinson

Available soon:

THE ETRUSCANS Christopher Smith
MICROECONOMICS Avinash Dixit
LANDSCAPE ARCHITECTURE
    Ian H. Thompson

THE EYE Michael Land
NUTRITION David A. Bender

For more information visit our website
www.oup.com/vsi/

Peter Atkins

PHYSICAL CHEMISTRY

A Very Short Introduction

OXFORD
UNIVERSITY PRESS

OXFORD
UNIVERSITY PRESS

Great Clarendon Street, Oxford, OX2 6DP,
United Kingdom

Oxford University Press is a department of the University of Oxford.
It furthers the University's objective of excellence in research, scholarship,
and education by publishing worldwide. Oxford is a registered trade mark of
Oxford University Press in the UK and in certain other countries

© Peter Atkins Limited 2014

The moral rights of the author have been asserted

First Edition published in 2014

Published in the United States of America by Oxford University Press
198 Madison Avenue, New York, NY 10016, United States of America

British Library Cataloguing in Publication Data
Data available

Library of Congress Control Number: 2014930362

ISBN 978-0-19-968909-5

Printed and bound by
CPI Group (UK) Ltd, Croydon, CR0 4YY

Preface

Physical chemistry provides the conceptual infrastructure of chemistry. It stands up to its knees in physics but has its head in inorganic and organic chemistry, the other two principal branches of chemistry. It follows that to understand modern chemistry, what it achieves, how it achieves it, and how it views the world, it is essential to understand the contributions of physical chemistry.

Physical chemistry, however, is built on a framework of mathematics. Therein lies the difficulty that aspiring students often find with the subject and which this author finds when seeking to share his vision of the subject without frightening his audience. With that in mind, I have sought to present an almost entirely verbal account of the subject so that readers can gather its attitudes and contributions to our understanding without being distracted by equations. Just occasionally, I show an equation to round out the discussion and show the basis of my verbalization; but I do that discreetly.

Standing in physics up to its knees necessarily means that physical chemistry draws on the wonderful contributions that physicists have made to our understanding of the world, and it will become apparent on reading these pages that Nobel Prizes for topics now fully part of physical chemistry have in large part been awarded to physicists. That is just a mark of the debt that physical chemistry owes to physics. I shall introduce these topics, which include aspects of quantum mechanics and thermodynamics, but only at the level I judge sufficient.

I hope these pages will reveal the contributions that physical chemistry has made to all branches of chemistry. It provides, or at least elucidates, elaborates, and justifies, much of the language of modern chemistry, and these pages will give some insight into that language and the concepts that all chemists use in their conversations and work. I hope too that they will illustrate the cultural contributions that the subject makes to our understanding of the natural world.

Finally, I must stress that although the principles of chemistry are, as far as we know, thoroughly understood, physical chemistry is still a vibrantly alive subject. Such is the sophistication of current and emerging instrumental techniques, including computation, that the sinews of physical chemistry are stretched to wring all the information from the data that the techniques provide. Moreover, new types of matter—I have in mind soft matter and nanosystems—and elaborate 'old' matter, biological matter, are now open to study by its techniques and are providing rich fields for its application. I have tried to identify these emerging fields in the concluding 'The current challenge' paragraphs of each chapter. Perhaps those paragraphs outline what might be regarded as the research interests of a hypothetical perfect research laboratory in modern physical chemistry, where collaboration with intellectually neighbouring disciplines should be the core strategy.

Peter Atkins
Oxford, 2013

Contents

List of illustrations

Chapter 1
Matter from the inside

One way to understand how a physical chemist thinks and contributes to chemistry is to start at the interior of an atom and then to travel out into the world of bulk matter. The interior of an atom is where much of the explanation of matter is to be found and it is here that a chemist is most indebted to physics. Within this realm, within an atom, explanations necessarily draw on quantum mechanics, that perplexing description of the behaviour of the very small. That quantum mechanics is central to their description should not be taken to be a warning that the rest of this chapter will be incomprehensible! I shall distil from that theory only the qualitative essence of what we need.

Atoms

The ancient Greeks speculated that matter was composed of atoms. That was pure speculation unsupported by any experimental evidence and so cannot be regarded as the beginning of physical chemistry. Experimental evidence for atoms was accumulated by John Dalton (1766–1844) in the very early 19th century when the use of the chemical balance allowed quantitative measurements to be made on the reactions that matter undergoes. Dalton inferred the existence of atoms from his measurements but had no way of assessing their actual sizes. He had no notion that nearly two centuries later, in the late 20th century, scientists would at last be able to see them.

For a physical chemist, an atom consists of a central, tiny, massive, positively charged *nucleus* surrounded by a cloud of much lighter, negatively charged *electrons*. Chemists have little interest in the details of the structure of the nucleus itself and are content to think of it as a tightly bound collection of two types of fundamental particle, positively charged *protons* and electrically neutral *neutrons*. The number of protons in the nucleus, the atom's 'atomic number', determines the identity of the element (1 for hydrogen, 2 for helium, and so on up to, currently, 118 for livermorium). The number of neutrons is approximately the same as the number of protons (none for ordinary hydrogen, 2 for ordinary helium, and about 170 for livermorium). This number is slightly variable, and gives rise to the different *isotopes* of the element. As far as a physical chemist is concerned, a nucleus is a largely permanent structure with three important properties: it accounts for most of the mass of the atom, it is positively charged, and in many cases it spins on its axis at a constant rate.

One particular nucleus will play an important role throughout this account: that of a hydrogen atom. The nucleus of the most common form of hydrogen is a single proton, a single ceaselessly spinning, positively charged fundamental particle. Although so simple, it is of the utmost importance in chemistry and central to the way that physical chemists think about atoms in general and some of the reactions in which they participate. There are two further isotopes of hydrogen: deuterium ('heavy hydrogen') has an additional neutron bound tightly to the proton, and tritium with two neutrons. They will play only a slight role in the rest of this account, but each has properties of technical interest to chemists.

The electronic structure of atoms

Physical chemists pay a great deal of attention to the electrons that surround the nucleus of an atom: it is here that the chemical action takes place and the element expresses its chemical personality. The principal point to remember in this connection is that the number of electrons in the atom is the same as the number of protons in the nucleus. The electric charges of electrons and protons are equal but opposite, so this equality of numbers ensures that the atom overall is electrically neutral. Thus, a hydrogen atom has a single electron around its nucleus, helium has two, livermorium has a crowded 118, and so on.

Quantum mechanics plays a central role in accounting for the arrangement of electrons around the nucleus. The early 'Bohr model' of the atom, which was proposed by Neils Bohr (1885–1962) in 1913, with electrons in orbits encircling the nucleus like miniature planets and widely used in popular depictions of atoms, is wrong in just about every respect—but it is hard to dislodge from the popular imagination. The quantum mechanical description of atoms acknowledges that an electron cannot be ascribed to a particular path around the nucleus, that the planetary 'orbits' of Bohr's theory simply don't exist, and that some electrons do not circulate around the nucleus at all.

Physical chemists base their understanding of the electronic structures of atoms on Schrödinger's model of the hydrogen atom, which was formulated in 1926. Erwin Schrödinger (1887–1961) was one of the founders of quantum mechanics, and in what he described as an episode of erotic passion whilst on vacation with one of his mistresses, he formulated the equation that bears his name and solved it for the location of the electron in a hydrogen atom. Instead of orbits, he found that the electron could adopt one of a variety of wave-like distributions around the nucleus, called *wavefunctions*, each wave corresponding to a particular energy.

Physical chemists adopt Schrödinger's solutions for hydrogen and adapt it as the starting point for their discussion of all atoms. That is one reason why the hydrogen atom is so central to their understanding of chemistry. They call the wave-like distributions of electrons *atomic orbitals*, suggesting a link to Bohr's orbits but indicating something less well-defined than an actual path.

We shall need some nomenclature. The lowest energy atomic orbital in a hydrogen atom is an example of an *s-orbital*. An electron in an s-orbital (physical chemists say 'in' when they mean having a distribution described by a particular orbital) can be pictured as a spherical cloud-like distribution that is densest at the nucleus and declines sharply with distance from it (Figure 1). That is, the electron is most likely to be found at the nucleus, and then progressively less likely to be found at points more distant from it. Incidentally, an electron in this orbital has no sense of rotation around the nucleus: it is effectively just hovering over and surrounding the nucleus. An atom is often said to be mostly empty

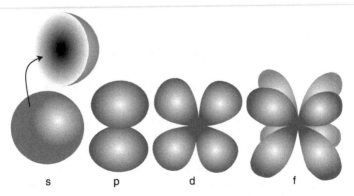

s   p   d   f

1. **The typical shapes of s-, p-, d-, and f-orbitals. The boundaries enclose regions where the electron is most likely to be found. The inset shows the internal structure of an s-orbital, with density of shading representing the probability of finding an electron at each location**

space. That is a remnant of Bohr's model in which a point-like electron circulates around the nucleus; in the Schrödinger model, there is no empty space, just a varying probability of finding the electron at a particular location.

There are other s-orbitals, each one of successively higher energy, and each one forming a spherical cloud-like shell at progressively greater distances from the nucleus. They are denoted 1s (the lowest energy s-orbital), 2s, 3s, and so on. In a hydrogen atom only the 1s orbital is occupied by the atom's lone electron.

There are also other wave-like solutions of the Schrödinger equation for the hydrogen atom (as shown in the illustration). What is called a *p-orbital* is like a cluster of two clouds on opposite sides of the nucleus, a *d-orbital* consists of four cloudlike regions, an *f-orbital* of eight such regions, and so on. There are two points to make in this connection: the s, p, d, f notation is derived from very early spectroscopic observations on atoms and its origin is no longer of any significance except to historians of science, but the notation persists and is a central part of every chemist's vocabulary. Second, the notation continues to g, h, etc.; but chemists almost never deal with these other orbitals and there is no need for us to consider them further.

Schrödinger went on to show that whereas there is only one way of wrapping a spherical shell around a nucleus (so there is only one s-orbital of any rank), there are three ways of wrapping a p-orbital around the nucleus (so there are three p-orbitals of any rank). Similarly, there are five ways of wrapping the even more complex d-orbitals around the nucleus (five d-orbitals of a given rank) and seven ways of wrapping f-orbitals. Moreover, for a given energy there are only certain types of orbitals that can exist.

The actual pattern is as follows:

1s
2s 2p
3s 3p 3d
4s 4p 4d 4f

and so on, with 1s having the lowest energy.

In a hydrogen atom, all the orbitals of the same rank (the same row in this list) have the same energy. In all other atoms, where there is more than one electron, the mutual repulsion between electrons modifies the energies and, broadly speaking, the list changes to

1s
2s 2p
3s 3p
4s 3d 4p

with more complicated changes in other orbitals. I shall make two points in this connection.

First, physical chemistry has a sub-discipline known as *computational chemistry*. As its name suggests, this sub-discipline uses computers to solve the very complex versions of the Schrödinger equation that arise in the treatment of atoms and molecules. I shall deal with molecules later; here I focus on the much simpler problem of individual atoms, which were attacked very early after the formulation of the Schrödinger equation by carrying out the enormously detailed calculations by hand. Now atoms are essentially a trivial problem for modern computers, and have been used to derive very detailed descriptions of the distribution of electrons in atoms and the energies of the orbitals. Nevertheless, although physical chemists can now calculate atomic properties with great accuracy almost at the touch of a button, they like to build up models of atoms that give insight into their structure and provide a sense of understanding rather than just a string of numbers. This understanding is then exported into inorganic and organic chemistry as well as other parts of physical chemistry. Models have been built of the way that repulsions between electrons in atoms other than hydrogen affect the order of orbital energies and the manner in which electrons occupy them.

The phrase 'the manner in which electrons occupy them' introduces another important principle from physics. In 1925 Wolfgang Pauli (1900–58) identified an important principle when confronted with some peculiar features of the spectroscopy of atoms: he noted that certain frequencies were absent in the spectra, and concluded that certain states of the atom were forbidden. Once quantum mechanics had been formulated it was realized that there is a deep way of expressing his principle, which we shall not use, and a much more direct way for our purposes, which is known as the *Pauli exclusion principle*:

> No more than two electrons may occupy any one orbital, and if two do occupy that orbital, they must spin in opposite directions.

We shall use this form of the principle, which is adequate for many applications in physical chemistry.

At its very simplest, the principle rules out all the electrons of an atom (other than atoms of one-electron hydrogen and two-electron helium) having all their electrons in the 1s-orbital. Lithium, for instance, has three electrons: two occupy the 1s orbital, but the third cannot join them, and must occupy the next higher-energy orbital, the 2s-orbital. With that point in mind, something rather wonderful becomes apparent: the structure of the Periodic Table of the elements unfolds, the principal icon of chemistry.

To see how that works, consider the first 11 elements, with 1 to 11 electrons (the numbers in brackets in this list):

H[1]   He[2]

Li[3]   Be[4]   B[5]   C[6]   N[7]   O[8]   F[9]   Ne[10]

Na[11]   ...

(If you need reminding about the names of the elements, refer to the Periodic Table in the Appendix at the end of this volume.) The first electron can enter the 1s-orbital, and helium's (He) second electron can join it. At that point, the orbital is full, and lithium's (Li) third electron must enter the next higher orbital, the 2s-orbital. The next electron, for beryllium (Be), can join it, but then it too is full. From that point on the next six electrons can enter in succession the three 2p-orbitals. After those six are present (at neon, Ne), all the 2p-orbitals are full and the eleventh electron, for sodium (Na), has to enter the 3s-orbital. Simultaneously, a new row of the Periodic Table begins. At a blow, you can now see why lithium and sodium are cousins and lie in the same column ('group') of the table: each one has a single electron in an s-orbital. Similar reasoning accounts for the entire structure of the Table, with elements in the same group all having analogous electron arrangements and each successive row ('period') corresponding to the next outermost shell of orbitals.

This is where physical chemistry segues into *inorganic chemistry*, the specific chemistry of all the elements. Physical chemistry has accounted for the general structure of the Periodic Table, and inorganic chemistry explores the consequences. It is quite remarkable, I think, that very simple ideas about the existence and energies of orbitals in alliance with a principle that governs their occupation accounts for the brotherhood of the elements, as represented by the Periodic Table.

The properties of atoms

Physical chemists do not quite let slip their grip on atoms at this point and hand them over to inorganic chemists. They continue to be interested in a variety of properties of atoms that stem from their electronic structure and which play a role in governing the chemical personalities of the elements.

Probably the most important property of an atom for the compounds that it can form is its size, specifically its *radius*. Although an atom has a fuzzy cloud of electrons around its nucleus, the density of the cloud falls off very rapidly at its edge—so speaking of the radius of the atom is not too misleading. In practice, atomic radii are determined experimentally by measuring the distance between two bonded atoms in a molecule or solid, and apportioning the distance to each atom in a prescribed way. The trends in atomic radius are found to correlate with the location of the element in the Periodic Table. Thus, on crossing the Table from left to right, atoms become smaller: even though they have progressively more electrons, the nuclear charge increases too, and draws the clouds in to itself. On descending a group, atoms become larger because in successive periods new outermost shells are started (as in going from lithium to sodium) and each new coating of cloud makes the atom bigger (Figure 2).

Second in importance is the *ionization energy*, the energy needed to remove one or more electrons from the atom. As we shall see, the ability of electrons to be partially or fully removed from an atom determines the types of chemical bonds that the atom can form and hence plays a major role in determining its properties. The ionization energy more or less follows the trend in atomic radii but in an opposite sense because the closer an electron lies to the positively charged nucleus, the harder it is to remove. Thus, ionization energy increases from left to right across the Table as the atoms become smaller. It decreases down a group because the outermost electron (the one that is most easily removed) is

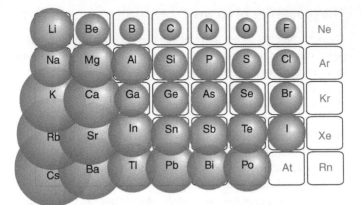

2. The variation in atomic radius across the Periodic Table. Radii typically decrease from left to right across a period and increase from top to bottom down a group. Only the 'main group' elements are shown here (not the transition elements)

progressively further from the nucleus. Elements on the left of the Periodic Table can lose one or more electrons reasonably easily: as we shall see in Chapter 4, a consequence is that these elements are metals. Those on the right of the Table are very reluctant to lose electrons and are not metals (they are 'non-metals').

Third in importance is the *electron affinity*, the energy released when an electron attaches to an atom. Electron affinities are highest on the right of the Table (near fluorine; ignore the special case of the noble gases). These relatively small atoms can accommodate an electron in their incompletely filled outer cloud layer and once there it can interact strongly and favourably with the nearby nucleus.

The importance of ionization energies and electron affinities becomes apparent when we consider the 'ions' that atoms are likely to form. An *ion* is an electrically charged atom. That charge comes about either because the neutral atom has lost one or more of its electrons, in which case it is a positively charged *cation* (pronounced 'cat ion') or because it has captured one or more electrons and has become a negatively charged *anion*. The names 'cation' and 'anion' were given to these ions by physical chemists studying the electrical conductivities of ions in solution, like salt dissolved in water, who noted that one class of ion moved 'up' an electrical potential difference and others moved 'down' it ('ion' comes from the Greek word for traveller, and 'an' and 'cat' are prefixes denoting 'up' and 'down', respectively). Elements on the left of the Periodic Table, with their low ionization energies, are likely to lose electrons and form cations; those on the right, with their high electron affinities, are likely to acquire electrons and form anions. This distinction brings us to the heart of one subject explored and elucidated by physical chemists: the nature of the chemical bond.

The ionic bond

A 'chemical bond' is what holds neighbouring atoms together to form the intricate structures of the world. All chemical bonds result from changes in the distribution of electrons in the bonded atoms, and so their formation falls very much into the domain of physical chemistry.

Chemists identify three types of bond: ionic, covalent, and metallic. I shall put off the discussion of the metallic bond until Chapter 4 and the discussion of metals. An ionically bonded substance consists of cations and anions clumped together as a result of the attraction between their opposite electric charges. The most famous exemplar of this type of bonding is sodium chloride, common salt, each grain of which consists of a myriad of sodium cations (Na^+) clumped together with an equal number of chloride ions (Cl^-). 'Clumping' is perhaps a misleading term, for the compound is not just a random jumble of ions but serried ranks of them, each Na^+ ion being surrounded by six Cl^- ions and each Cl^- ion likewise surrounded by six Na^+ ions in a highly orderly array that extends throughout the little crystal (Figure 3).

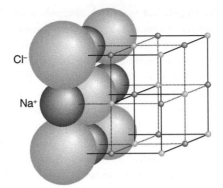

3. The structure of sodium chloride (NaCl). Each sodium ion (Na⁺) is surrounded by six chloride ions (Cl⁻), and each Cl⁻ ion is surrounded by six Na⁺ ions. This pattern is repeated throughout the crystal

Physical chemists identify several significant features of the ionic bond. One is that there is not a discrete ion–ion bond. What I have called 'clumping together' is a result of all the ions in the crystal interacting with each other: a Na⁺ ion interacts favourably with the Cl⁻ ions that surround it, but is repelled by the Na⁺ ions that surround each of those Cl⁻ ions; in turn it interacts favourably with the next rank of Cl⁻ ions, unfavourably with the next rank of Na⁺ ions, and so on. The attractions and repulsions diminish with distance, but nevertheless ionic bonding should be thought of as a global, not local, aspect of the crystal.

The strength of an ionic bond, and the reason it exists, is that the ions in the crystal have a lower energy than the same number of widely separated sodium and chlorine atoms. This is where ionization energies and electron affinities come into play to determine whether bond formation is favourable. Here, physical chemists turn into profit-and-loss accountants. At first sight the balance sheet shows loss rather than profit. The ionization energy of sodium is low, but its single outermost electron doesn't simply fall off: there is a substantial energy investment needed to remove it. That investment is partly recovered by the electron affinity of chlorine, with energy released when the electron attaches to a chlorine atom to form Cl^-. However, that overall process still requires energy because the energy lowering that occurs in the second step is much less than the energy required in the first ionization step. The additional release of energy that turns loss into profit is the very strong net favourable interaction between the ions that are formed, and which is released as they clump together. In effect, this huge lowering of energy drags the ions into existence and results in a stable crystal of sodium chloride.

You can now see a third characteristic of ionic bonds: that they form primarily between atoms on the left and right of the Periodic Table. Atoms on the left have low ionization energies, so the energy investment for forming cations is reasonably low. Atoms on the right have reasonably high electron affinities, so there is a reasonable lowering of energy when they form anions. This net loss (it is always a loss, for ionization energies are high and electron affinities not very helpful) is turned into profit by the resulting global net attraction between the ions.

The covalent bond

Elements on the right of the Periodic Table have such high ionization energies that if they alone are involved in bond formation, then loss is never turned into profit. Physical chemists have identified another way in which atoms can form liaisons with one another, especially if like elements towards the right of the Table they are reluctant to release electrons and form cations. These atoms compromise: they *share* their electrons and form what is known as a *covalent bond*.

I need to step back a little from this development before explaining the concept and consequences of covalent bonding, and make a couple of points. First, the significance of the 'valent' part of covalent. This term is derived from the Latin word for 'strength' ('*Valete*!', 'Be strong!', was the Roman farewell); the word *valence* is now a common chemical term in chemistry referring to the theory of bond formation. The 'co' part is an indication of cooperation between atoms for achieving bond strength.

Second, it is hard to pinpoint the moment in history when physical chemistry became an identifiable discipline within chemistry. There were certainly physical chemists in the 19th century: Michael Faraday among them and Robert Boyle (Chapter 4) even earlier in the 17th century, even though they did not use the term. However, one branch of physical chemistry did emerge right at the beginning of the 20th century when the electron (discovered in 1897) became central to chemical explanation, and perhaps a good starting point for this branch of the discipline is the identification of the covalent bond. That identification is due to the American chemist Gilbert Lewis (1875–1946), who proposed in 1916 that the covalent bond is a shared pair of electrons. It is one of the scandals of intellectual history that Lewis never received the Nobel Prize despite his several seminal contributions to chemistry and physical chemistry in particular.

As I have indicated, a covalent bond is a shared pair of electrons. If an atom has too high an ionization energy for cation formation to be energetically feasible, it might be content to surrender partial control of an electron. Moreover, it might recover some of that small energy investment by accommodating a share in an electron supplied by another like-minded atom provided that overall there is a lowering of energy. I have purposely used anthropomorphic terms, 'content to' and 'like-minded', for they commonly creep into chemists' conversations as shorthand for what they really mean, which is an allusion to the changes in energy that accompany the redistribution of electrons. I shall avoid them in future, for they are sloppy and inappropriate (but often whimsically engaging, and like much conversational informality, help to avoid pedantic circumlocutions).

The principal characteristic of covalent bonding is that, in contrast to ionic bonding, it is a *local* phenomenon. That is, because the electrons are shared between neighbours, an identifiable bond exists between those two neighbours and is not distributed over myriad ions. One consequence of this local character is that covalently bonded species are discrete molecules, like hydrogen, H_2 (H—H), water, H_2O (H—$\ddot{O}$—H), and carbon dioxide, CO_2 (O=C=O). I use these three molecules as examples for a particular reason.

First, H_2 reveals that atoms of the same element might bond covalently together. Such forms of the elements are common among the non-metals (think of oxygen, O_2, and chlorine, Cl_2). The formula H—H also shows how chemists denote covalent bonds: by a dash between the linked atoms. Each such dash indicated a shared pair of electrons. The water molecule is H_2O; I have included it both to show that not all the electrons possessed by an atom need be involved in covalent bond formation and that by forming two bonds the oxygen atom is surrounded by eight electrons (four pairs). The unused pairs are called *lone pairs*. The fact that eight electrons are present indicates that the oxygen atom has gone on to form as many bonds as it can until it has completed what we have already seen as the capacity of its outermost cloud of electrons. This so-called 'octet formation' is one of the guiding but not always reliable principles of bond formation. Finally, I have included carbon dioxide because it shows that more than one pair of electrons can be shared between two atoms: the = indicates a 'double bond', consisting of two shared pairs of electrons. There are also molecules with triple bonds (three shared pairs), and very occasionally quadruple bonds.

Not all covalently bonded species are discrete molecules. It is possible in some instances for covalent bond formation to result in extended networks of bonds. The most famous example is diamond, where each carbon atom is singly-bonded to four neighbours, and each of those neighbours bonded to their neighbours, and so on throughout the crystal. I shall return to a discussion of these structures in Chapter 4.

The quantum mechanics of bonds

Although he developed several speculative fantasies, Lewis had no idea of the real reason why an electron pair is so central to covalent bond formation. That understanding had to await the development of quantum mechanics and its almost immediate application to valence theory. That green shoot of an application has since evolved into a major branch of physical chemistry namely *theoretical chemistry* or, more specifically for those dealing with numerical calculations, *computational chemistry*.

Physical chemists, initially in collaboration with physicists, have developed two quantum mechanical theories of bond formation, *valence-bond theory (VB theory)* and *molecular orbital theory (MO theory)*. The former has rather fallen out of favour but its language has left an indelible mark on chemistry and chemists still use many of its terms. MO theory has swept the field as it has proved to be much more readily implemented on computers. Because VB theory language is still used throughout chemistry but calculations are carried out using MO theory, all chemists, under the guidance of physical chemists, need to be familiar with both theories.

VB theory was formulated by the physicists Walter Heitler (1904–81), Fritz London (1900–54), and John Slater (1900–76) and elaborated by the chemist Linus Pauling (1901–93) almost as soon as quantum mechanics had been born, in 1927. This collaboration of physicists and chemists is an excellent example of the foundation that underlies physical chemistry: it also illustrates the intellectual fecundity of youth, as all its originators were in their 20s. The core idea of VB theory is that a wavefunction can be written for the pair of electrons that form a bond. The reason for the importance of the pair being that each electron has a spin (as I mentioned when discussing atoms) and for that bond wavefunction to exist the two electrons must spin in opposite directions. Note that this description focuses on neighbouring pairs of atoms: one atom provides an electron that spins clockwise, the other atoms provides an electron that spins counterclockwise, and the two electrons pair together. Close analysis of the resulting wavefunction shows that this pairing allows the two electrons to accumulate between the two nuclei and, through their electrostatic attraction for the nuclei, effectively glue the two atoms together.

The theory quickly ran into initial difficulties with carbon and its fundamental compound, methane, CH_4. This is one place where Pauling made his principal contribution and where the language he introduced survives to pervade modern chemistry. The lowest energy state of a carbon atom has four electrons in its outermost cloud (its 'valence shell'), two of which are already paired. The remaining two can pair with electrons provided by two hydrogen atoms, but that results in CH_2, not CH_4. Pauling proposed the idea of *promotion*, in which it is envisaged that there is a notional energy investment in transferring one of the paired 2s electrons into an empty 2p-orbital. Now all four electrons can be available for pairing with electrons providing by hydrogen atoms, so allowing the formation of CH_4.

Promotion, however, raises another problem, for one of the bonds (the one involving the original 2s-orbital) is different from the other three (formed by pairing with electrons in the three 2p-orbitals), but in CH_4 all four bonds are known to be the same. Pauling overcame this problem by proposing the concept of *hybridization*. Each 2s- and 2p-orbital is like a wave centred on the nucleus: like all waves they can be considered to interfere if they occupy the same region of space, which in this atom they do. Analysis of the interference between the four waves shows that they result in four interference patterns that correspond to orbitals that are identical except for pointing in different directions (to the corners of a tetrahedron in this case). These 'hybrid orbitals', Pauling proposed, should be used to form the covalent bonds of the CH_4 molecule. The result is four identical bonds and a molecule that looks like a miniature tetrahedron, exactly as is observed.

Another simple molecule, hydrogen chloride (HCl) presented another problem, for analysis of the VB wavefunction showed that it was a poor description of the distribution of electrons in one important respect: it never allowed both electrons to be on the chlorine atom at the same time. We have seen that chlorine, on the right of the Periodic Table, has a high electron affinity, so it is physically implausible that the two electrons of the bond don't spend a good proportion of their time close to it. Here Pauling proposed the concept of *resonance*. Instead of writing a single wavefunction representing an H–Cl molecule, another one should be written for a tiny (and in this case, local) ionic structure, H^+Cl^-, and both should contribute to the actual structure of the molecule. This so-called 'superposition' of wavefunctions, allowing both to contribute to the description, is called resonance, and its inclusion improves the description of the molecule (and results in a lower calculated energy: always a sign of improvement).

Resonance was simultaneously the saviour and the cause of the downfall of VB theory. It saved the day for small molecules, where there are only a few resonance structures to consider, but it proved an insurmountable barrier for big molecules where thousands of structures could contribute to the resonance.

MO theory was initially formulated by Robert Mulliken (1896–1986) and Friedrich Hund (1896–1997) in 1927 as a rival to VB theory; the name was coined by Mulliken in 1932. It can be regarded as a natural extension of the theory of the electronic structure of atoms. As we have seen, in an atom electrons occupy (that is, have distributions described by) wavefunctions called atomic orbitals. In a molecule, according to MO theory, electrons occupy wavefunctions called 'molecular orbitals' that spread over all the nuclei present in the molecule and help bind them together into a stable arrangement. According to the Pauli exclusion principle, each molecular orbital, like an atomic orbital, can accommodate no more than two electrons and their spins must be paired, so immediately the importance of Lewis's electron pairing is explained.

The Schrödinger equation for molecular orbitals is far too difficult to solve, so approximations are essential. First, molecular orbitals are constructed from all the atomic orbitals that are present in the molecule. Thus, in H_2, the two 1s atomic orbitals are used. As for any waves, interference occurs where these wavefunctions spread into the same region of space (as mentioned in the discussion of hybridization, but here it is interference between waves centred on neighbouring atoms). In this case, the interference may either enhance the amplitude of the waves where they overlap or diminish it. The former gives rise to a *bonding orbital*, for electrons that occupy it will be found accumulated between the two nuclei and glue them together by the attraction between opposite charges. Destructive overlap results in a reduction in amplitude of the wavefunction between the two nuclei and thus removes the electrons from where it is beneficial for them to be. The combination is therefore called an *antibonding orbital*, as electrons in it tend to drive the nuclei apart (Figure 4).

Although the Schrödinger equation is too difficult to solve for molecules, powerful computational procedures have been developed by theoretical chemists to arrive at numerical solutions of great accuracy. All the procedures start out by building molecular orbitals from the available atomic orbitals and then setting about finding the best formulations. This branch of physical chemistry is the subject of intense development as the power of computers increases. Depictions of electron distributions in molecules are now commonplace and very helpful for understanding the properties of molecules. It is particularly relevant to the development of new pharmacologically active

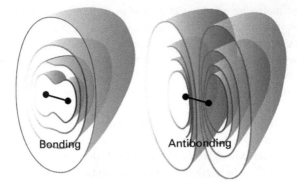

4. The bonding and antibonding orbitals in a two-atom molecule. In a bonding orbital, the electrons accumulate between the two nuclei and bind them together. In an antibonding orbital, they are excluded from the internuclear region and the two nuclei push each other apart. Similar comments apply to molecules composed of many atoms. The surfaces are contours of equal probability for finding the electron

drugs, where electron distributions play a central role in determining how one molecule binds to another and perhaps blocks the deleterious function of a dangerously invasive molecule.

These computational procedures fall into three broad categories. The *semi-empirical* procedures acknowledge that some computations are just too time-consuming (and therefore expensive) unless some of the parameters needed in the calculations are taken from experiment. The purer *ab initio* ('from scratch') procedures stand aloof from this compromise, and seek to perform the calculation with no input other than the identities of the atoms present. Initially only small molecules could be treated in this way, but as computational power increases so larger and larger molecules are captured in its net. One currently fashionable procedure is a blend of these approaches: *density functional theory* is widely applicable as it provides a swift and reasonably accurate way of calculating the properties of a wide range of molecules.

The current challenge

Bond formation is essentially fully understood apart from some recondite questions relating to the role of the contributions of potential and kinetic energy and to the presence of relativistic effects in heavy atoms. The principal focus now is on the efficient and reliable computation of the electronic structures of ever bigger molecules, including biologically interesting molecules, the electronic properties of nanostructures, and detailed descriptions of the behaviour of electrons in molecules. Chemical reactions are increasingly being modelled computationally, with insight being gained into how bonds are broken and new ones formed. Drug discovery, the identification of pharmacologically active species by computation rather than *in vivo* experiment, is an important target of modern computational chemistry.

Chapter 2
Matter from the outside

In the early days of physical chemistry, maybe back in the 17th century but more realistically during the 19th century, its practitioners, lacking the extraordinary instruments we now possess, investigated appearances rather than burrowing into the then unknown inner structure of matter. That activity, when performed quantitatively rather than merely reporting on the way things looked, proved extraordinarily fruitful, especially in the field of study that came to be known as 'thermodynamics'.

Thermodynamics is the science of energy and the transformations that it can undergo. It arose from considerations of the efficiencies of steam engines in the hands of engineers and physicists who were intent on extracting as much motive power as possible from a pound of coal or, as electricity developed, a pound of zinc. Those early pioneers had no conception, I suspect, that their discoveries would be imported into chemistry so effectively that thermodynamics would come to play a central role in understanding chemical reactions and acquire a reach far beyond what the phrase 'transformation of energy' suggests.

Modern chemical thermodynamics is still all about energy, but in the course of establishing the laws that govern energy it turns out that sometimes unsuspected relations are discovered between different properties of bulk matter. The usefulness of that discovery is that measurements of one property can be used to determine another that might not be readily accessible to measurement.

There are currently four laws of thermodynamics, which are slightly capriciously labelled 0, 1, 2, and 3. The Zeroth Law establishes the concept of temperature, and although temperature is of vital importance for the discussion of all forms of matter and their properties, I accept that its conceptual and logical basis is not of much concern to the general run of physical chemists and shall not discuss it further. (I touch on the significance of temperature again in Chapter 3.) The First Law concerns the hugely important conservation of energy, the Second Law that wonderfully illuminating property entropy, and the Third Law the seemingly frustrating inaccessibility of the absolute zero of temperature. As no chemistry happens at absolute zero, the Third Law might seem not to have much relevance to physical chemists, but in fact it plays an important role in the way they use data and I cannot ignore it. It is in the domain of the Second Law that most of the relations between measurable properties are found, and I shall explain what is involved in due course.

'Matter from the outside' is an appropriate title of a chapter on thermodynamics, for in its purest form, which is known as 'classical thermodynamics', no discussion or deduction draws on any model of the internal constitution of matter. Even if you don't believe in atoms and molecules, you can be an effective (if blinkered) thermodynamicist. All the relations derived in classical thermodynamics refer to observable properties of bulk matter and do not draw on the properties of the constituent atoms and molecules. However, much insight into the origin of these bulk properties and the relations between them is obtained if the blinkers are removed and the atomic constitution of matter acknowledged. That is the role of the fearsomely named *statistical thermodynamics*, to which I give an introduction in Chapter 3.

Throughout this chapter I shall focus on the applications of thermodynamics in physical chemistry. It goes without saying that thermodynamics is also widely applicable in engineering and physics, where it originated. It is also a largely mathematical subject, especially when it comes to establishing relations between properties (and even defining those properties), but I shall do my best to present its content verbally.

The First Law

As far as physical chemists are concerned, the First Law is an elaboration of the *law of the conservation of energy*, the statement that energy can be neither created nor destroyed. The elaboration is that thermodynamics includes the transfer of energy as heat whereas the dynamics of Newton and his descendants, does not. The central concept of the First Law is the *internal energy*, U, the total energy of whatever region of the world we are interested in, the 'system', but not including its energy due to external causes, such as the motion of the entire system through space. The detailed experiments of James Joule (1818–89) in the middle of the 19th century established that the internal energy of a system could be changed either by doing work or by heating the system. Work, a concept from dynamics, involves moving against an opposing force; heat, the new concept in thermodynamics, is the transfer of energy that makes use of a temperature difference. The failure of often well-meaning but also innumerable charlatan- and greed-driven attempts to create a perpetual motion machine, a machine that can generate work without any input of energy (for instance, as heat) finally led to the conclusion that *the internal energy of a system that is isolated from external influences does not change*. That is the First Law of thermodynamics.

I need to interject a remark here. A system possesses energy, it does not possess work or heat (even if it is hot). Work and heat are two different modes for the *transfer* of energy into or out of a

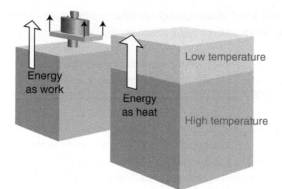

5. Work is a mode of transfer that is equivalent to raising a weight; heat is a mode of transfer that results from a temperature difference. These two modes of transfer decrease or increase the internal energy of a system

system. Work is a mode of transfer that is equivalent to raising a weight; heat is a mode of transfer that results from on a temperature difference (Figure 5). When we say 'a system is hot', we don't mean that it has a lot of heat: it has a high temperature on account of the energy it currently stores.

For practical reasons, in many instances physical chemists are less concerned with the internal energy than with a closely related property called the *enthalpy*, H. The name derives from the Greek words for 'heat inside'; although there is no 'heat' as such inside a system; the name is quite sensible, as I shall shortly explain. For completeness, although it is immaterial to this verbal discussion, if you know the internal energy of a system, then you can calculate its enthalpy simply by adding to U the product of pressure and volume of the system ($H = U + pV$). The significance of the enthalpy—and this is the important point—is that a change in its value is equal to the output of energy as heat that can be obtained from the system provided it is kept at constant pressure. For instance, if the enthalpy of a system falls by 100 joules when it undergoes a certain change (such as a chemical reaction), then we

25

know that 100 joules of energy can be extracted as heat from the system, provided the pressure is constant.

The relation between changes in enthalpy and heat output lies at the heart of the branch of physical chemistry, and more specifically of chemical thermodynamics, known as *thermochemistry*, the study of the heat transactions that accompany chemical reactions. This aspect of chemistry is of vital importance wherever fuels are deployed, where fuels include not only the gasoline of internal combustion engines but also the foods that power organisms. In the latter application, the role of energy deployment in organisms is studied in *bioenergetics*, the biochemically specific sub-section of thermochemistry.

The principal instrument used by thermochemists is a *calorimeter*, essentially a bucket fitted with a thermometer, but which, like so much scientific apparatus, has been refined considerably into an elaborate instrument of high precision subject to computer control and analysis. Broadly speaking, the reaction of interest is allowed to take place in the bucket and the resulting temperature rise is noted. That temperature rise (in rare cases, fall) is then converted to a heat output by comparing it with a known reaction or the effect of electrical heating. If the calorimeter is open to the atmosphere (and therefore at constant pressure), then the heat output is equal to the enthalpy change of the reaction mixture. Even if the reaction vessel is sealed and its contents undergo a change in pressure, chemists have ways of converting the data to constant-pressure conditions, and can deduce the enthalpy change from the observed change in temperature.

Thermochemists, by drawing on the First Law of thermodynamics know, with some relief, that they don't have to examine every conceivable reaction. It may be that the reaction of interest can be thought of as taking place through a sequence of steps that have already been investigated. According to the First Law, the enthalpy change for the direct route must be the same as the sum of the enthalpy changes along the indirect route (just as two paths up a mountain between the same two points must lead to the same change in altitude), for otherwise you would have a neat way of creating energy: start with a compound, change it; change it back to the original compound by a different path, and extract the difference in energy. Therefore, they can tabulate data that can be used to predict enthalpy changes for any reaction of interest. For instance, if the enthalpy changes for the reactions A → B and B → C are known, then that of the reaction A → C is their sum. (Physical chemists, to the friendly scorn of inorganic and organic chemists, often don't commit themselves to specific reactions. They argue that it is to retain generality that they use A, B, and C; their suspicious colleagues often wryly suspect that such generality actually conceals ignorance.)

In the old days of physical chemistry (well into the 20th century), the enthalpy changes were commonly estimated by noting which bonds are broken in the reactants and which are formed to make the products, so A → B might be the bond-breaking step and B → C the new bond-formation step, each with enthalpy changes calculated from knowledge of the strengths of the old and new bonds. That procedure, while often a useful rule of thumb, often gave wildly inaccurate results because bonds are sensitive entities with strengths that depend on the identities and locations of the other atoms present in molecules. Computation now plays a central role: it is now routine to be able to calculate the difference in energy between the products and reactants, especially if the molecules are isolated as a gas, and that difference easily converted to a change of enthalpy. The computation is less reliable for reactions that take place as liquids or in solution and progress there is still being made.

Enthalpy changes are very important for a rational discussion of changes in physical state (vaporization and freezing, for instance), and I shall return to them in Chapter 5. They are also essential for applications of the Second Law.

The Second Law

The Second Law of thermodynamics is absolutely central to the application of thermodynamics in chemistry, often appearing in disguised form, but always lurking beneath the surface of the discussion of chemical reactions and their applications in biology and technology.

The First Law circumscribes possible changes (they must conserve the total amount of energy, but allow for it to be shipped from place to place or transferred as heat or work). The Second Law identifies from among those possible changes the changes that can occur spontaneously. The everyday meaning of 'spontaneously' is 'without external agency'; in physical chemistry it means without having to do *work* to bring the change about. It does not, as the word is sometimes coloured in common discourse, mean fast. Spontaneous changes can be very slow: the name simply indicates that they have a *tendency* to take place without intervention. The conversion of diamonds to graphite is spontaneous in this sense, but it is so slow that the change can be disregarded in practice. The expansion of a gas into a vacuum is both spontaneous and fast. Its compression into a smaller volume is not spontaneous: we have to do work and push on a piston to bring it about. In short: the First Law identifies possible changes, the Second Law identifies from among those possible changes the ones that are spontaneous.

The Second Law identifies spontaneous changes as those accompanied by an increase in *entropy*. Entropy, S, is a measure of the quality of energy: low entropy corresponds to high quality in a sense I shall explain; high entropy corresponds to low quality. The name is derived from the Greek words for 'inward turning', which gives an impression that it is concerned with changes taking place within a system. The increase in entropy therefore corresponds to the decline of the quality of energy, such as it becoming more widely dispersed and less useful for doing work.

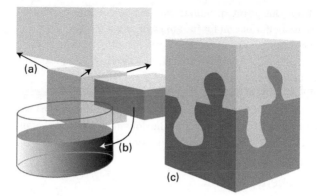

6. Entropy is a measure of 'disorder'. Thus, it increases (a) when a gas disperses, (b) when a solid melts, and (c) when two substances mingle

Entropy is commonly associated as a measure of 'disorder', such as the dispersal of energy or the spreading of a gas (Figure 6). That is a very helpful interpretation to keep in mind. The quantitative expression for calculating the numerical value of the change in entropy was first proposed by the German physicist Rudolph Clausius (1822–88) in 1854: his formula implies that to calculate the entropy change we need to note the energy transferred carefully as heat during the change and divide it by the absolute temperature at which the change occurs. Thus, if 100 joules of heat enters a beaker of water at 20°C (293 K), the entropy of the water increases by 0.34 joules per kelvin. A very different but complementary and hugely insightful approach was developed by Ludwig Boltzmann, and I treat it in Chapter 3.

Physical chemists have used Clausius's formula to compile tables of entropies of a wide range of substances (this is where they also have to use the Third Law, as I shall explain shortly), and can use them to assess the entropy change when reactants with one entropy change into products with a different entropy. This is essential information for judging whether a reaction is spontaneous at any given temperature (not necessarily fast, remember; just spontaneous). Here is another example of the utmost importance of the reliance of chemists on the achievements of physicists, in this case Clausius, for understanding why one reaction 'goes' but another doesn't is central to chemistry and lies in the domain of physical chemistry, the interface of physics and chemistry.

Free energy

There is a catch. In all presentation of chemical thermodynamics it is emphasized that to use the Second Law it is necessary to consider the total entropy change, the sum of the changes within the system (the reaction mixture) and its surroundings. The former can be deduced from the tables of entropies that have been compiled, but how is the latter calculated?

This is where enthalpy comes back to play a role. If we know the enthalpy change taking place during a reaction, then provided the process takes place at constant pressure we know how much energy is released as heat into the surroundings. If we divide that heat transfer by the temperature, then we get the associated entropy change in the surroundings. Thus, if the enthalpy of a system falls by 100 joules during a reaction taking place at 25°C (298 K), that 100 joules leaves the system as heat and enters the surroundings and consequently their entropy increases by 0.34 joules per kelvin. All we have to do is to add the two entropy changes, that of the system and that of the surroundings, together, and identify whether the total change is positive (the reaction is spontaneous) or negative (the reaction is not spontaneous).

Physical chemists have found a smart way to do this fiddly part of the calculation, the assessment of the entropy change of the surroundings. Well, to be truthful, it was another theoretical physicist, the American Josiah Gibbs (1839–1903), who in the 1870s brought thermodynamics to the stage where it could be regarded as an aspect of physical chemistry and indeed effectively founded this branch of the subject. Once again, we see that physics provides the springboard for developments in physical chemistry.

Gibbs noted that the two changes in entropy, that of the system and that of the surroundings, could, provided the pressure and temperature are constant, be calculated by dealing solely in changes of what we now call the *Gibbs energy*, *G*, of the system alone and apparently ignoring the surroundings entirely. Remember how the enthalpy is calculated by adding pV to the internal energy ($H = U + pV$); similarly the Gibbs energy of a system is calculated simply by subtracting the product of the temperature and entropy from the enthalpy: $G = H - TS$. For instance, the entropy of 100 mL of water at 20°C (293 K) is 0.388 kilojoules per kelvin, so its Gibbs energy is about 114 kilojoules less than whatever its enthalpy happens to be.

Actual numbers, though, are not particularly interesting for our purposes: much more important is the presence of the negative sign in the definition $G = H - TS$. When this expression is analysed it turns out that, provided the pressure and temperature are constant, a spontaneous change corresponds to a *decrease* in Gibbs energy. This makes the Gibbs energy of the system alone a signpost of spontaneous change: all we need do is to compile tables of Gibbs energies for tables of enthalpies and entropies, work out how *G* changes in a reaction, and note whether the change is negative. If it is negative, then the reaction is spontaneous (Figure 7).

I need to make another interjection. Although it might seem to be more natural to think of a process as being natural if it corresponds to 'down' in some property, in this case the Gibbs energy, it must be remembered that the Gibbs energy is a

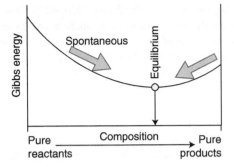

7. A reaction is spontaneous in the direction of decreasing Gibbs energy. When the Gibbs energy is a minimum, the reaction is at equilibrium with no tendency to change in either direction

disguised version of the total entropy. The spontaneous direction of change is invariably 'up' in total entropy. Gibbs showed chemists that, at the price of losing generality and dealing only with changes at constant temperature and pressure, they could deal with a property of the system alone, the Gibbs energy, and simply by virtue of the way it is defined with that negative sign, 'down' in Gibbs energy is actually 'up' in total entropy.

There is a very important aspect of spontaneity that I now need to introduce. Chemists are very interested in the composition at which a chemical reaction has no further tendency to change: this is the state of *chemical equilibrium*. Neither the forward reaction nor its reverse is spontaneous when the mixture has reached equilibrium. In other words, at the composition corresponding to equilibrium, no change in composition, either the formation of more product or their decomposition, corresponds to a decrease in Gibbs energy: the Gibbs energy has reached a minimum, with a change in either direction corresponding to an increase in G. All a physical chemist needs to do to predict the composition at equilibrium is to identify the composition at which the Gibbs energy has reached a minimum. There are straightforward ways of doing that calculation, and the equilibrium composition of virtually any reaction at any temperature can be calculated provided the Gibbs energy data are available.

The ability to predict the equilibrium composition of a reaction and how it depends on the conditions is of immense importance in industry, for it is pointless to build a chemical factory if the yield of product is negligible, and questions of economy draw on finding the conditions of temperature and pressure that are likely to optimize the yield.

Finally, I need to explain the term 'free energy' that I have used as the title of this section and thereby open another important door on the biological and technological applications of the Second Law.

There are two kinds of work. One kind is the work of expansion that occurs when a reaction generates a gas and pushes back the atmosphere (perhaps by pressing out a piston). That type of work is called 'expansion work'. However, a chemical reaction might do work other than by pushing out a piston or pushing back the atmosphere. For instance, it might do work by driving electrons through an electric circuit connected to a motor. This type of work is called 'non-expansion work'. Now for the crucial point: a change in the Gibbs energy of a system at constant temperature and pressure is equal to the maximum non-expansion work that can be done by the reaction. Thus, a reaction for which the Gibbs energy decreases by 100 joules can do up to 100 joules of non-expansion work. That non-expansion work might be the electrical work of driving electrons through an external circuit. Thus, we arrive at a crucial link between thermodynamics and *electrochemistry*, which includes the generation of electricity by chemical reactions and specifically the operation of electric batteries and fuel cells (see Chapter 5).

There is another crucial link that this connection reveals: the link of thermodynamics with biology is that one chemical reaction might do the non-expansion work of building a protein from amino acids. Thus, a knowledge of the Gibbs energies changes accompanying metabolic processes is very important in bioenergetics, and much more important than knowing the enthalpy changes alone (which merely indicate a reaction's ability to keep us warm). This connection is a major contribution of physical chemistry to biochemistry and biology in general.

The Third Law

I have cheated a little, but only in order to make fast progress with the very important Second Law. What I glossed over was the measurement of entropy. I mentioned that changes in entropy are measured by monitoring the heat supplied to a sample and noting the temperature. But that gives only the value of a *change* in entropy. What do we take for the initial value?

The Third Law of thermodynamics plays a role. Like the other two laws, there are ways of expressing the Third Law either in terms of direct observations or in terms of the thermodynamic properties I have mentioned, such as the internal energy or the entropy. One statement of the Third Law is of the first kind: it asserts that *the absolute zero of temperature cannot be reached in a finite number of steps*. That version is very important in the field of cryogenics, the attainment of very low temperatures, but has little directly to do with physical chemistry. The version that does have direct implications for physical chemistry is logically equivalent (it can be shown), but apparently quite different: *the entropies of all perfect crystals are the same at the absolute zero of temperature*. For convenience (and for reasons developed in Chapter 3) that common value is taken as zero.

Now we have a starting point for the measurement of absolute (or 'Third-Law') entropies. We lower the temperature of a substance to as close as possible to absolute zero, and measure the heat supplied at that temperature. We raise the temperature a little, and do the same. That sequence is continued until we reach the temperature at which we want to report the entropy, which is simply the sum of all those 'heat divided by temperature' quantities. This procedure still gives the change in entropy between zero temperature and the temperature of interest, but the former entropy is zero, by the Third Law, so the change gives us the absolute value of the entropy too.

Thus, the Third Law plays a role in physical chemistry by allowing us to compile tables of absolute entropies and to use those values to calculate Gibbs energies. It is a crucial part of the fabric of chemical thermodynamics, but is more of a technical detail than a contribution to insight. Or is that true? Can it be that the Third Law does provide insight? We shall see in Chapter 3 how it may be interpreted.

Relations between properties

I remarked at the beginning of this chapter that chemical thermodynamics provides relations, sometimes unexpected, between the properties of substances, those properties often being apparently unrelated to considerations of the transformations of energy. Thus, measurements of several properties might be stitched together using the guidance of thermodynamics to arrive at the value of a property that might be difficult to determine directly.

This section is difficult for me to generate not only because I want to avoid mathematics but also because you would be unlikely to find many of the relations in the least interesting and be unmoved by knowing that there are clever ways of measuring seemingly recondite properties! The best I can come up with to illustrate the kind of relation involved is that between different types of a moderately familiar property, heat capacity (commonly referred to as 'specific heats').

Let me establish the problem. A heat capacity, C, is the ratio of the heat supplied to a substance to the temperature rise produced. For instance, if 100 joules of energy is supplied as heat to 100 mL of water, its temperature rises by 0.24°C, so its heat capacity is 420 joules per degree. But when physical chemists wear their pernickety accountant's hat, they stop to think. Did the water expand? If it did, then not all the energy supplied as heat remained in the sample because some was used to push back the atmosphere. On the other hand, suppose the water filled a rigid, sealed vessel and the same quantity of energy as heat was supplied to it. Now no expansion occurs, no work was done, and all that energy remains inside the water. In this case its temperature rises further than in the first case so the heat capacity is smaller (the heat supplied is the same but the temperature rise is greater, so their ratio is smaller). In other words, when doing precise calculations (and thermodynamics is all about circumspection and precision), we have to decide whether we are dealing with the heat capacity at constant pressure (when expansion is allowed) or at constant volume (when expansion is not allowed). The two heat capacities, which are denoted C_p and C_V, respectively, are different.

What thermodynamics, and specifically a combination of the First and Second Laws, enables us to do, is to establish the relation between C_p and C_V. I shall not give the final relation, which involves the compressibility (how its volume changes with pressure) and its expansivity (how its volume changes with temperature), but simply want to mention that by using these properties, which can be measure in separate experiments, C_p and C_V can be related. There is one special case of the relation, however, that is worth presenting. For a perfect gas (a concept I develop in Chapter 4, but essentially an idealized gas in which all interactions between the molecules can be ignored, so they are flying around freely), the relation between the two heat capacities is $C_p - C_V = Nk$, where N is the total number of molecules in the sample and k is the fundamental constant known as Boltzmann's constant (once again, see Chapter 3). Because Nk is positive, we can infer that C_p is greater than C_V, just as our discussion suggested should be the case.

This example might seem rather trivial and not particularly interesting (as I warned). However, it does reveal that physical chemists can deploy the laws of thermodynamics, laws relating to matter from the outside, to establish relations between properties and to make important connections.

The current challenge

Thermodynamics was developed for bulk matter. Attention now is shifting towards special forms of matter where 'bulk' might not be applicable: to very tiny nanosystems and to biological systems with their special functions and structures. One branch of thermodynamics that has made little progress since its formulation is 'irreversible thermodynamics', where the focus is on the rate at which energy is dissipated and entropy is generatedby systems that are not at equilibrium: a Fifth Law might be lurking currently undiscovered there.

Chapter 3
Bridging matter

I have adopted the eyes of a physical chemist to lead you on an exploration of matter from the inside in Chapter 1, where the focus was on the properties of individual atoms and molecules. I have used them again to show how a physical chemist explores it from the outside in Chapter 2, where classical thermodynamics is aloof from any knowledge about atoms and molecules. There is, of course, a link between these notional insides and outsides, and the branch of physical chemistry known as *statistical thermodynamics* (or *statistical mechanics*) forms that bridge.

At this level of presentation, some will think that it is a bridge too far, for statistical thermodynamics, as its faintly fearsome name suggests, is very mathematical. Nevertheless, it is a central duty of science to show how the world of appearances emerges from the underworld of atoms, and I would omit a large chunk of physical chemistry if I were to evade this topic. I shall keep the discussion qualitative and verbal, but you should know that lying beneath the qualitative surface there is a rigid infrastructure of mathematics. Because I am stripping away the mathematics, and the soul of statistical thermodynamics is almost wholly mathematical, this chapter is mercifully short! You could easily skip it, but in doing so you would miss appreciating how some physical chemists think about the properties of matter.

The central point about statistical thermodynamics is that it identifies the bulk properties of a sample with the *average* behaviour of all the molecules that constitute it. There is no need, it claims, to work out the behaviour of each molecule, just as there is no need in certain sociological settings to follow the behaviour of an individual in a big enough sample of the population. Yes, there will be idiosyncrasies in molecular behaviour just as there are in human behaviour, but these idiosyncrasies, these 'fluctuations', are negligible in the vast collections of molecules that make up typical samples. Thus, to calculate the pressure exerted by a gas, it is not necessary to consider the impact of each molecule on the walls of a container: it is sufficient (and highly accurate given the large numbers involved, with trillions and trillions of molecules in even very small samples) to deal with the steady swarming impact of myriad molecules and to ignore the tiny fluctuations in the average impact as the storm abates or surges.

The Boltzmann distribution

There is one extraordinary conclusion from statistical thermodynamics that is absolutely central to physical chemistry and which accounts for great swathes of the properties of matter, including its reactions as well as illuminating that elusive property, temperature. If there is one concept to take away from this chapter, this is it.

One consequence of quantum mechanics is that an atom or molecule can exist only certain energy states; that is, its energy is 'quantized' and cannot be varied continuously at will. At the absolute zero of temperature, all the molecules are in their lowest energy state, their 'ground state'. As the temperature is raised, some molecules escape from the ground state and are found in higher energy states. At any given temperature above absolute zero, such as room temperature, there is a spread of molecules over all their available states, with most in the ground state and progressively fewer in states of increasing energy.

The distribution is ceaselessly fluctuating as molecules exchange energy, perhaps as collisions in a gas, when one partner in the collision speeds off at high speed and the other is brought nearly to a standstill. It is impossible, given the many trillions of molecules in a typical sample, to keep track of the fluctuating distribution. However, it turns out that in such typical samples there is one overwhelmingly most probable distribution and that all other distributions can be disregarded: this dominant distribution is the *Boltzmann distribution*.

I will put a little more visual flesh on to this description. Suppose you had a set of bookshelves, with each shelf representing an energy level, and you tossed books at the shelves at random. You could imagine that in one set of tries you arrived at a particular distribution, maybe one with equal numbers on each of the shelves. Achieving that distribution is possible but highly unlikely. In another try, half the books land on the lowest shelf and the other half all on the top shelf. That, again is unlikely, but much more likely than the first distribution. (In fact, if you had 100 books and 10 shelves, then it would be about 2×10^{63} times more likely!) As you went on hurling books, noting the distribution on each try, you would typically get different distributions each time. However, one distribution would turn up time and time again: that most probable distribution is the Boltzmann distribution.

Comment

So as not to complicate this discussion, I am leaving out a crucial aspect: that each distribution might correspond to a different total energy; any distribution that does not correspond to the actual energy of the system must be discarded. So, for instance, unless the temperature is zero, we reject the distribution in which all the books end up on the lowest shelf. All on the top shelf is ruled out too. Physical chemists know how to do it properly.

The Boltzmann distribution takes its name from Ludwig Boltzmann (1844–1906), the principal founder of statistical thermodynamics. Mathematically it has a very simple form: it is an exponentially decaying function of the energy (an expression of the form $e^{-E/kT}$, with E the energy, T the absolute temperature, and k the fundamental constant we now call Boltzmann's constant). That means that the probability that a molecule will be found in a state of particular energy falls off rapidly with increasing energy, so most molecules will be found in states of low energy and very few will be found in states of high energy.

Comment

The Boltzmann distribution is so important (and simple) that I need to show a version of it. If two states have the energies E_1 and E_2, then at a temperature T the ratio of the numbers of molecules in those two states is $N_2 / N_1 = e^{-(E_2 - E_1)/kT}$

The reach of the tail of the decaying exponential function, its overall shape in fact, depends on a single parameter that occurs in the expression: the temperature, T (Figure 8). If the temperature is low, then the distribution declines so rapidly that only the very lowest levels are significantly populated. If the temperature is high, then the distribution falls off very slowly with increasing energy, and many high-energy states are populated. If the temperature is zero, the distribution has all the molecules in the ground state. If the temperature is infinite, all available states are equally populated. This is the interpretation of temperature that I mentioned: *it is the single, universal parameter that determines the most probable distribution of molecules over the available states.* This too is the beginning of the elucidation of thermodynamics through Boltzmann's eyes, for temperature is the parameter introduced (as I indicated in Chapter 2 without

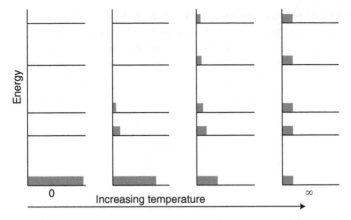

8. The Boltzmann distribution for a series of temperatures. The horizontal lines depict the allowed energies of the system and the tinted rectangles the relative numbers of molecules in each level at the stated temperature

going into details) by the Zeroth Law. We shall see below that Boltzmann's insight also elucidates the molecular basis of the First, Second, and Third Laws.

The Boltzmann distribution captures two aspects of chemistry: stability and reactivity. At normal temperatures, the distribution implies that most molecules are in states of low energy. That corresponds to stability, and we are surrounded by molecules and solids that survive for long periods. On the other hand, some molecules do exist in states of high energy, especially if the temperature is high, and these are molecules that undergo reactions. Immediately we can see why most chemical reactions go more quickly at high temperatures. Cooking, for instance, can be regarded as a process in which, by raising the temperature, the Boltzmann distribution reaches up to higher energies, more molecules are promoted to states of high energy, and are able to undergo change. Boltzmann, short-sighted as he was, saw further into matter than most of his contemporaries.

Molecular thermodynamics

There is much more to statistical thermodynamics than the Boltzmann distribution, although that distribution is central to just about all its applications. Here I want to show how physical chemists draw on structural data, such as bond lengths and strengths, to account for bulk thermodynamic properties: this is where spectroscopy meets thermodynamics.

One of the central concepts of thermodynamics, the property introduced by the First Law, is the internal energy, U, of a system. As I pointed out in Chapter 2, the internal energy is the total energy of the system. As you can therefore suspect, that total energy can be calculated if we know the energy state available to each molecule and the number of molecules that are in that state at a given temperature. The latter is given by the Boltzmann distribution. The former, the energy levels open to the molecules, can be obtained from spectroscopy or by computation. By combining the two, the internal energy can be calculated, so forging the first link between spectroscopy or computation and thermodynamics.

The second link, a link to the Second Law, draws on another proposal made by Boltzmann. As we saw in Chapter 2, the Second Law introduces the concept of entropy, S. Boltzmann proposed a very simple formula (which in fact is carved on his tomb as his epitaph) that enables the entropy to be calculated effectively from his distribution. Therefore, because knowledge of the internal energy and the entropy is sufficient to calculate all the thermodynamic properties, from Boltzmann's distribution a physical chemist can calculate all those properties from information from spectroscopy or the results of computation.

I am glossing over one point and holding in reserve another. The glossed-over point is that the calculations are straightforward only if all the molecules in the sample can be regarded as independent of one another. Physical chemists currently spend a great deal of time (their own and computer time) on tackling the problem of calculating and thereby understanding the properties of interacting molecules, such as those in liquids. This remains a very elusive problem and, for instance, the properties of that most important and common liquid, water, remain largely out of their grasp.

The point I have held in reserve is much more positive. Boltzmann's formula for calculating the entropy gives great insight into the significance of this important property and is the origin of the common interpretation of entropy as a measure of disorder. When they consider entropy and its role in chemistry, most physical chemists think of entropy in this way (although some dispute the interpretation, focusing on the slippery nature of the word 'disorder'). Thus, it is easy to understand why the entropy of a perfect crystal of any substance is zero at the absolute zero of temperature (in accord with the Third Law) because there is no positional disorder, or uncertainty in location of the atoms, in a perfect crystal. Furthermore, all the molecules, as we have seen, are in their ground state, so there is no 'thermal disorder', or uncertainty in which energy states are occupied.

Likewise, it is easy to understand why the entropy of a substance increases as the temperature is raised, because it becomes increasingly uncertain which state a molecule will be in as more states become thermally accessible. That is, the thermal disorder increases with temperature. It is also easy to understand, I think, why the entropy of a substance when its solid form melts and increases even more when its liquid form vaporizes, for the positional disorder increases, especially in the second case.

As we saw in Chapter 2, in thermodynamics, the direction of spontaneous change is that in which the entropy (the total entropy, that of the surroundings as well as the system of interest) increases. Through Boltzmann's eyes we can now see that in every case the direction of spontaneous change, the 'arrow of time', is the direction in which disorder is increasing. The universe is simply crumbling, but such is the interconnectedness of events that the crumbling generates local artefacts of great complexity, such as you and me.

Molecular reactions

Physical chemists use statistical thermodynamics to understand the composition of chemical reaction mixtures that have reached equilibrium. It should be recalled from Chapter 2 that chemical reactions seem to come to a halt before all the reactants have been consumed. That is, they reach equilibrium, a condition in which forward and reverse reactions are continuing, but at matching rates so there is not net change in composition. What is going on? How do physical chemists think about equilibrium on a molecular scale?

In Chapter 2, I explained that physical chemists identify equilibrium by looking for the composition at which the Gibbs energy has reached a minimum. That, at constant temperature and pressure, is just another way of saying the composition at which any further change or its reverse results in a decrease in the total entropy, so it is unnatural. Because we have seen that statistical thermodynamics can account for the entropies of substances by drawing on structural data, we should expect to use similar reasoning to relate equilibrium compositions to structural data, such as that obtained from spectroscopy or computation.

The molecular reason why chemical reactions reach equilibrium is very interesting (I think) and is exposed by statistical thermodynamics. Let's take a very simple, typical physical chemistry, reaction, in which $A \rightarrow B$. We can use statistical thermodynamics to calculate the enthalpy and entropy of both the reactant A and the product B, and therefore we can use these values to calculate their Gibbs energies from structural data. Now imagine that all the A molecules are pinned down and can't migrate from their starting positions. If the A molecules gradually change into B molecules the total Gibbs energy of the system slides in a straight line that reflects the abundances of the two species, from its value for 'only A' to the value for 'only B'. If the Gibbs energy of B is less than that of A, then the reaction will go to completion, because 'pure B' has the lowest Gibbs energy of all possible stages of the reaction. This will be true of all reactions of the same form, so at this stage we should expect all such $A \rightarrow B$ reactions to go to perfect completion, with all A converted to B. They don't.

What have we forgotten? Remember that we said that all the A were pinned down. That is not true in practice, for as the A convert into B, the A and B mingle and at any intermediate stage of a reaction there is a haphazardly mingled mixture of reactants and products. Mixing adds disorder and increases the entropy of the system and therefore lowers the Gibbs energy (remember that $G = H - TS$ and its minus sign, so increasing S reduces G). The greatest increase in entropy due to mixing, and therefore the greatest negative contribution to the Gibbs energy, occurs when A and B are equally abundant. When this mixing contribution is taken into account, it results in a minimum in the total Gibbs energy at a composition intermediate between pure A and pure B, and that minimum corresponds to the composition at which the

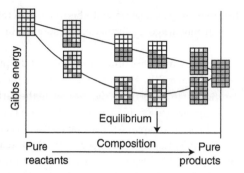

Gibbs energy

Equilibrium ↓

Pure reactants — Composition → Pure products

9. The role of mixing in the determination of the equilibrium composition of a reaction mixture. In the absence of mixing, a reaction goes to completion; when mixing of reactants and products is taken into account, equilibrium is reached when both are present

reaction is at equilibrium (Figure 9). The Gibbs energy of mixing can be calculated very easily, and we have already seen that the Gibbs energy change between pure A and pure B can be calculated from structural data, so we have a way of accounting for the equilibrium composition from structural data.

Similar reasoning is used by physical chemists to account for the effect of changing conditions on the equilibrium composition. As I remarked in Chapter 2, industrial processes, in so far as they are allowed to reach equilibrium (many are not), depend on the optimum choice of conditions. Statistical thermodynamics, through the Boltzmann distribution and its dependence on temperature, allows physical chemists to understand why in some cases the equilibrium shifts towards reactants (which is usually unwanted) or towards products (which is normally wanted) as the temperature is raised. A rule of thumb, but little insight, is provided by a principle formulated by Henri Le Chatelier (1850–1936) in 1884, that a system at equilibrium responds to a disturbance by tending to oppose its effect. Thus, if a reaction releases energy as heat (is 'exothermic'), then raising the temperature will oppose the formation of more products; if the

Bridging matter

47

reaction absorbs energy as heat (is 'endothermic'), then raising the temperature will encourage the formation of more product.

An implication of Le Chatelier's principle was one of the problems confronting the chemist Fritz Haber (1868–1934) and the chemical engineer Carl Bosch (1874–1940) in their search in the early 20th century for an economically viable synthesis of ammonia (which I treat in more detail in Chapter 6). They knew that, because the reaction is exothermic, raising the temperature of their reaction mixture of nitrogen and hydrogen opposed the formation of ammonia, which plainly they did not want. That realization forced them to search for a catalyst that would facilitate their process at moderate rather than very high temperatures.

Statistical thermodynamics provides an explanation of Le Chatelier's principle. (I remain focused on its role in explaining the effect of temperature: other influences can apply and be

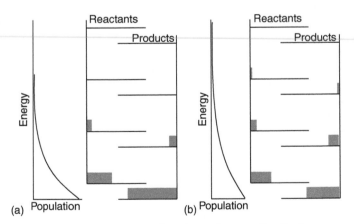

10. (a) At a given temperature there is a Boltzmann distribution over the states of the reactants and products; (b) when the temperature is raised in this exothermic reaction, more of the higher-energy reactant states are occupied, corresponding to a shift in equilibrium in favour of the reactants

explained.) If a reaction is exothermic, then the energy levels of the products lie lower than those of the reactants. At equilibrium there is a Boltzmann distribution of molecules over all the available states, and if a molecule is in a 'reactant' state then it is present as a reactant, and if it is in 'product' state then it is present as a product. When the temperature is raised the Boltzmann distribution stretches up further into the high-energy reactant states, so more molecules become present as reactants (Figure 10), exactly as Le Chatelier's principle predicts. Similar reasoning applies to an endothermic reaction, when the 'product' states lie above the 'reactant' states and become more populated as the temperature is raised, corresponding to the encouragement of the formation of products.

A statistical perspective

I did warn that by stripping the mathematics out of the presentation statistical thermodynamics it might seem more vapour than substance. However, I hope it has become clear that this branch of physical chemistry is truly a bridge between molecular and bulk properties and therefore an essential part of the structure of physical chemistry.

In the first place, statistical thermodynamics unites the two great rivers of discussion, on the one hand the river of quantum mechanics and its role in controlling the structures of individual molecules, and on the other hand the river of classical thermodynamics and its role in deploying the properties of bulk matter. Thermodynamics, as I have stressed, can in principle stand aloof from any molecular interpretation, but it is immeasurably enriched when the properties it works with are interpreted in terms of the behaviour of molecules. Moreover, a physical chemist would typically regard understanding as incomplete unless a property of bulk matter (which includes the chemical reactions in which it participates) has been interpreted in molecular terms.

Second, physical chemistry is a quantitative branch of chemistry (not the only one, of course, but one in which quantitative argument and numerical conclusions are paramount), and statistical thermodynamics is right at its quantitative core, for although general arguments can be made about anticipated changes in energy and entropy, statistical thermodynamics can render those anticipations quantitative.

The current challenge

Statistical thermodynamics is difficult to apply in all except simple model systems, such as systems consisting of independent molecules. It is making brave attempts to conquer systems where interactions are of considerable and sometimes overwhelming importance, such as the properties of liquids (that enigmatic liquid water, especially) and of solutions of ions in water. Even common salt dissolved in water remains an elusive system. New frontiers are also opening up as very special systems come into the range of chemistry, such as biological molecules and nanoparticles. In the latter, questions arise about the relevance of large-scale statistical arguments, which are appropriate for bulk matter, to the very tiny systems characteristic of nanoscience.

Chapter 4
States of matter

A physical chemist is interested in all the 'states' of matter, the physical form in which it is found. There are three traditional states: gas, liquid, and solid. There are well-known borderline states, such as the liquid crystals that are used on many visual displays, which are half-liquid and half-solid. A new arrival in the solid category consists of nanomaterials, matter present on a scale of about 100 nm and less. Another new state is almost pure surface in the form of 'graphene'. I shall say more about these new arrivals later. Some people regard 'plasma', a gas of ions, as a different state of matter, but currently it has little relevance to mainstream chemistry and I shall ignore it.

Gas

A physical chemist is very much at home in a gas. This is the easiest state of matter to describe and to make quantitative predictions about. Indeed, the study of gases could be taken to be yet another springboard of physical chemistry, with Robert Boyle (1627–91) and his investigation of the 'spring of the air' in Oxford in about 1660. The study of gases was stimulated further by another feature that was to persist through the centuries, when mankind took to the air in balloons and technological advances depended on further basic research. I have in mind Jacques Charles (1746–1823), his flight in

1783, and the later formulation of his law by another pioneering chemical balloonist, Joseph Louis Gay-Lussac (1778–1850), in 1802.

Boyle, Charles, and Gay-Lussac jointly identified the bulk properties that characterize what we know as a *perfect gas* (or 'ideal gas'), essentially the soul of gassiness. Their experiments were crude and the subtle nuances that distinguish one gas from another were washed away in the imprecision of their measurements. That was a good thing at that primitive epoch of science, for they were able to identify simple laws without being distracted by small discrepancies that could be accommodated later. Thus, Boyle established the relation between the volume and the pressure of a gas ($V \propto 1/p$) and Charles and Gay-Lussac the relation between the volume and what we now know as the absolute temperature ($V \propto T$). To this behaviour in the early 19th century Amedeo Avogadro (1776–1856) added a final contribution, when he hypothesized that equal volumes of gases at the same temperature and pressure contain equal numbers of molecules. In other words, the volume of a gas is proportional to the number of molecules present ($V \propto N$). Thus, by combining these observations, was established the 'perfect gas law', $V \propto NT/p$. Rather astonishingly, the missing universal constant of proportionality in this relation, universal in the sense that it is the same constant for all gases regardless of their identity, is Boltzmann's constant, k, which I introduced in Chapter 3, and the law is then $V = NkT/p$. That k occurs is another sign of the underlying ubiquity of the Boltzmann distribution in the description of matter, and I shall explore why it arises in this context shortly.

Comment

The relation $V = NkT/p$ is normally written in the form $pV = nRT$. The 'gas constant' R is related to k by $R = N_A k$, where N_A is Avogadro's constant and n is the amount of substance. I am reluctant to smother the discussion with definitions, except discreetly.

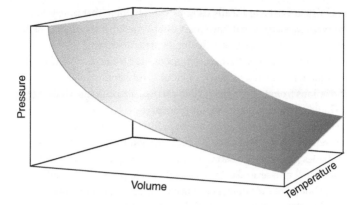

11. The perfect gas equation of state summarizes the allowed states of a perfect gas, a gas in which the molecules neither attract nor repel each other. This surface shows the values of pressure allowed for given values of the volume and temperature of the gas

The perfect gas law, which I will now rearrange into $p = NkT/V$ by multiplying both sides by p/V, is an example of an *equation of state*, an expression that shows how the pressure is related to other variables under our control, in this case the number of molecules in the sample, the temperature, and the volume it occupies (Figure 11). The perfect gas equation of state is the simplest example and is found to be obeyed by actual gases increasingly well as the pressure is reduced. That is, the perfect gas law is a *limiting law* in the sense that it is strictly valid only in a certain limit, in this case as $p \rightarrow 0$. Actual gases at nonzero pressures are described by more complicated equations of state which are not always known but for which approximate expressions have been established. Happily, though, the perfect gas equation is a very good approximation at normal pressures, and is widely used as a starting point in discussions in physical chemistry.

The existence of the perfect gas equation of state brings me to a very important aspect of physical chemistry: model building. Physical chemists regard empirically established relations (that is, relations between properties that have been established by experiment, such as an equation of state) as a challenge: what molecular interpretation can be devised that accounts for the observed law or relation and then how can it be modified to accommodate deviations from the simple law? In the case of gases, the challenge is to build a model of a perfect gas that agrees with the observed perfect gas equation of state and then to refine the model so that the equations of state of actual gases are obtained. Model building pervades physical chemistry, not just in the description of gases, and I shall give several examples as these chapters unfold. Indeed, some hold that the whole of science is based on building models of physical reality; much of physical chemistry certainly is.

The model of matter that accounts for the perfect gas equation of state is one of the most extraordinary in science: it is essentially ignorance. The *kinetic theory of gases* accepts that we really don't know anything about the details of what is happening in a gas except that the molecules are in ceaseless random motion and undergo collisions with one another and the walls of the container. When that model is explored, it turns out that $p \propto N/V$, with a constant of proportionality related to the average speed of the molecules in the sample. That mean speed can be calculated from the Boltzmann distribution, which (as I explained in Chapter 3) also arises from ignorance, ignorance about the instantaneous distribution of molecules over their available energy states, in this case the energy states associated with their free motion through space. The missing constant turns out to be simply kT, so we have arrived at the expression $p = NkT/V$, which is exactly the perfect gas equation of state. From ignorance about the details of what the molecules are doing, we have arrived at an experimentally verified conclusion.

I need to make several points about this model and the conclusions drawn from it. First, it ignores the fact that real molecules interact with one another, attracting each other when they are nearby and repelling one another as soon as they come into contact. Because this contribution has been ignored we can suspect that the deviations that actual gases (so-called 'real gases') show from the perfect gas equation of state are due to these interactions. Therefore, to enhance the model and account for deviations from perfect behaviour, these interactions must be incorporated. That is one of the tasks of statistical thermodynamics (Chapter 3) or, if that approach is too challenging, by suggesting ways in which the perfect gas equation of state might be modified. For instance, Johannes van der Waals (1837–1923) proposed a modification in 1873 in which the attractions and repulsions were modelled by two parameters, a and b. One of these parameters, b, is easy to understand: it represents the repulsions that forbid one molecule to occupy the space already occupied by another, and simply reduces the volume available to the gas; thus, the repulsions change $p = NkT/V$ to $p = NkT/(V - b)$. The resulting *van der Waals equation of state* (which also includes a term proportional to a for the effect of the attractions between molecules) captures many of the properties of real gases and is commonly used by physical chemists to model their behaviour (Figure 12).

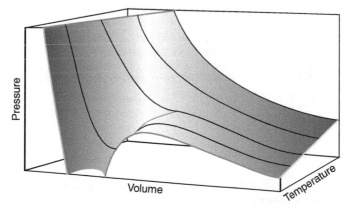

12. **When the gas molecules do interact with one another, the pressure depends on the volume in a more complicated way than is shown in Figure 11**

A second point is that the model can be developed to deduce some details about the average behaviour of molecules in a gas and so to provide some insight into what is going on in any sample of gas. Insight like that is crucial for understanding the properties of gases and the reactions they undergo. Thus, the Boltzmann distribution can be used to calculate the average speed of the molecules and the fraction of molecules having a specified speed (Figure 13). For reasonably light molecules (such as the major constituents of air, N_2 and O_2) at room temperature, the molecules are whizzing around at an average speed of about 500 m/s (about 1000 mph). That speed is consistent with what we know about the propagation of sound, the speed of which is about 340 m/s through air: for sound to propagate, molecules must adjust their position to give a wave of undulating pressure, so the rate at which they do so must be comparable to their average speeds.

The kinetic model can be elaborated to predict the average frequency with which molecules collide and the average

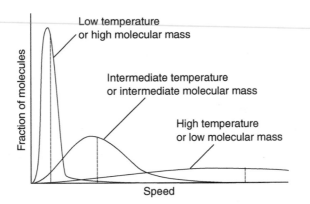

13. The 'Maxwell–Boltzmann distribution of speeds' shows the fraction of molecules having a given speed at a specified temperature and can be used to calculate the mean speed of the molecules. Light molecules not only have high average speeds (marked by the dotted lines) but also a high proportion move very rapidly

distance they travel between collisions. To do so, the pointlike molecules of the basic theory are imagined as being surrounded by a sphere of a certain radius that depends on the identity of the molecules, and counting as a 'hit' whenever two such spheres come into contact. When this elaboration is worked through, it turns out that a typical N_2 or O_2 molecule in air makes a collision every nanosecond and travels about 1000 molecular diameters between collisions. To put this scale into perspective: if a molecule is thought of as being the size of a tennis ball, then it travels about the length of a tennis court between collisions. Each molecule makes about a billion collisions a second.

Physical chemists, in collaboration with physicists, have used the kinetic model to discuss the rate at which properties are transported through gases and the rate of impacts of molecules on surfaces. For instance, they are able to relate the *thermal conductivity* of a gas, its ability to transport heat down a temperature gradient, to molecular properties. The model is also central to the discussion of the rates at which chemical reactions take place in gases (Chapter 6). It really is quite extraordinary how much information can be extracted from almost complete ignorance.

Liquid

Liquids are a hard nut to crack. Physical chemists have to be interested in them because so much chemistry takes place in them as a medium that it would be a grave omission not to explore and model their properties. The difficulty with understanding them is that they are intermediate between two simple extremes: the chaotic turmoil of molecules in a gas and the rigid arrays of atoms in a solid. Moreover, the most essential liquid to understand, water, is so bizarre that a deep understanding of its properties remains tantalizingly elusive.

There are several aspects of liquids that are open to experimental investigation; they include where the molecules lie at any instant, how fast they are moving, and how easy it is for them to migrate past one another. They can also be studied by monitoring the migration of dissolved substances through them and by observing the way in which they absorb different kinds and colours of electromagnetic radiation. Each experimental method, as is typical of physical chemistry, invites the construction of a model to explain the results. Unlike in a gas, of course, where intermolecular interactions are, to a first approximation, negligible, in liquids, where one molecule jostles against another, they are central to the description.

Even the concept of where molecules lie is problematical, for they are in ceaseless motion, migrating past one another, twisting and turning as they go. To describe the structure, such as it is, of a liquid all we can do is to report the probability that one molecule will be at a certain distance from another. Information of that kind can be obtained experimentally by much the same techniques as are used for solids (X-ray diffraction; see below), and it is found that a typical molecule is surrounded by a shell of other molecules, and then a less well-defined shell surrounds those, and maybe another even more ill-defined shell around those (Figure 14). Water, as in so many respects, is peculiar, as the liquid is found to consist of regions of moderately well-defined 'ice-like' regions amid amorphous 'liquid-like' regions. The starting point for modelling liquids is to think about how small spheres stack together when shaken together, and the general features of the local structures of liquids are quite well emulated by pots loosely packed with ball bearings.

Physical chemists have several ways of studying the dynamics of molecules in liquids, the characteristics of their motion. The picture to bear in mind is that of a jostling crowd of people; if

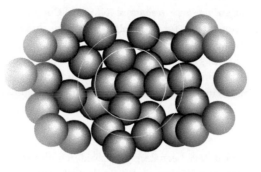

14. A depiction of the structure of a liquid composed of spherical molecules. Each molecule is surrounded by a shell of other molecules, and then a less well-defined shell surrounds those, and maybe another even more ill-defined shell around those. The whole structure is mobile

liquid flow is being studied, then the picture turns into that of a crowd of spectators leaving a packed stadium.

The classical way of studying the motion of molecules is to measure its 'viscosity', its resistance to flow, over a range of temperatures. Viscosities, though, are notoriously difficult to interpret although of considerable practical significance, as in the design of chemical plant where liquids are forced to flow through pipes from place to place. Insight into the motion of molecules themselves is obtained from a variety of techniques. One is *neutron scattering*: streams of the uncharged cousin of the proton, neutrons, are directed into a liquid sample and bounce off the liquid's molecules with a variety of energies. By examining those energies a picture can be constructed of the dynamics of the molecules and clusters of cooperating molecules acting in a coordinated manner can be constructed. That other helpful particle, the photon, shot from lasers can also give information of a similar kind.

I discuss *nuclear magnetic resonance* (NMR) in Chapter 7, but need to mention it here, for it is another technique that can be used to study the motion of molecules in liquids. In particular it can be used to study how molecules rotate in the dense, restrictive environment within a liquid. It is found that in some cases molecules rotate in a series of tiny steps, taking many steps to rotate completely, whereas in some liquids a molecule might swing through a right angle or so at every step.

Physical chemists have been deeply involved in building models of substances dissolved in water and using those dissolved substances to examine the dynamics of molecular motion. Ionic solutes (think sodium chloride in water, with its Na^+ and Cl^- ions dispersed among the water molecules) are particularly helpful in this respect because the ions can be dragged through the solvent by applying electric fields and the rate at which they move examined by monitoring the resulting electric current. Some initially counterintuitive results are obtained. Thus, it is found that big ions migrate more readily than little ions. That observation led to a picture in which the ions were regarded as being surrounded by water molecules that stuck to them and made them appear bigger because the whole assembly had to migrate through the liquid. Little ions attract water molecules more strongly than big ions (they give rise to a stronger electric field), so with their coating of water molecules masquerade as bigger ions.

This model of ions moving through water was nearly demolished when it was found that hydrogen ions, the smallest ions of all, far from behaving like little ions and moving slowly with a great coating of water molecules, were the most nimble of all! The interpretation gave insight into the behaviour of protons in water, for it was concluded that they didn't actually move at all. Instead, there is just an effective migration of a proton as neighbouring water molecules shift their ownership of their protons. Thus, in a chain of water molecules like

$$\begin{array}{ccccc} \overset{\displaystyle H}{|} & & \overset{\displaystyle H}{|} & & \overset{\displaystyle H}{|} \\ H\overset{+}{-}O-H- & --O-H- & --O-H \end{array}$$

a subtle shift in ownership results in

$$\begin{array}{ccccc} \overset{\displaystyle H}{|} & & \overset{\displaystyle H}{|} & & \overset{\displaystyle H}{|} \\ H-O- & ---H-O- & --H-O\overset{+}{-}H \end{array}$$

and the proton effectively and suddenly pops up at the other end of the chain and there is no actual lumbering migration of the original proton.

An early model of the structure of solutions of ions in water is still in use today by physical chemists as a first approximation to the behaviour of ions in solution and in particular their thermodynamic properties. This model was proposed by Peter Debye (1884–1966) and Erich Hückel (1896–1980) in 1923 and, to my mind, is an excellent example of the process of building models as it involves a lot of small steps, each of which is guided by physical insight. The core idea is that a cation (a positive ion) in solution tends to be surrounded by anions (negative ions), and vice versa. Thus, if we think in terms of the time average of their environments, each ion is surrounded by a flimsy 'ionic atmosphere' of opposite charge. The interaction of the central ion with its oppositely charged atmosphere lowers its energy and so makes it less thermodynamically active.

Debye and Hückel developed this model quantitatively and deduced what has become known as the *Debye–Hückel law* in which the energy lowering is proportional to the square root of the concentration of ions present. In fact, theirs is another example of the limiting laws characteristic of physical chemistry: in this case, the approximations made in the course of its development mean that it is strictly valid only in the limit of zero concentration. Despite that limitation, the law is a good starting point for more sophisticated theories of ionic solutions.

Solid

Physical chemistry is on much firmer ground with solids. Here, instead of motion being paramount (as in gases), motion constrained but important (as in liquids), to a first approximation there is no motion at all for all the atoms of a solid can be regarded as in fixed positions. Motion does survive in solids, but it is largely just the quivering of molecules in place as they vibrate, and initially can be ignored.

The principal experimental technique for the determination of the structures of solids is *X-ray diffraction* (or 'X-ray crystallography'). However, I have to admit that so central is an understanding of the arrangement of atoms in molecules to molecular biology and inorganic chemistry, that molecular biologists and inorganic chemists probably regard X-ray diffraction as their own; indeed, I think it true to say that physical chemists on the whole are happy to let slip the technique from their intellectual grasp but keep a tight grip on its conclusions. Their conclusions overlap with the interests of materials scientists, for here, as in biology, structure is allied closely with function.

That—the distance that physical chemists largely keep from the technique itself—being so, I will not dwell on it for more than a sentence or so, and focus instead on what it reveals. In short, X-ray diffraction makes use of the fact that electromagnetic radiation (which includes X-rays) consists of waves that can interfere with one another and give rise to regions of enhanced and diminished intensity. This so-called 'diffraction pattern' is characteristic of the object in the path of the rays, and mathematical procedures can be used to interpret the pattern in terms of the object's structure. Diffraction occurs when the wavelength of the radiation is comparable to the dimensions of the object. X-rays have wavelengths comparable to the separation of atoms in solids, so are ideal for investigating their arrangement.

Crystals, as was speculated long ago by René Haüy (1743–1822) in about 1784, following similar speculations by Johannes Kepler in 1611 and Robert Hooke in 1665, and confirmed by X-ray diffraction in the 20th century, consist of serried ranks of atoms in uniform arrays. The most primitive model for accounting for the structures of crystals of elemental metals (silver, copper, iron, etc.) is to regard each atom as a hard sphere and to consider how these spheres may be stacked together in layers to reproduce the entire crystal. Many metals have 'close-packed' structures in which the spheres pack together in the most efficient way possible (Figure 15). One consequence is that metals are dense materials as so much space is occupied by their closely packed atoms.

Ionic solids present a double problem: cations and anions have different radii and opposite charges. The lowest energy arrangement is obtained only if cations are surrounded by anions, and vice versa (the sodium chloride structure shown in Figure 3

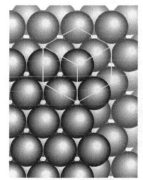 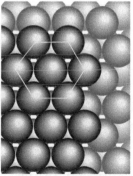

Cubic close packing   Hexagonal close packing

15. There are two ways of packing identical uncharged spheres together, one gives a 'cubic' arrangement and the other gives a 'hexagonal' arrangement. Copper is an example of the former, zinc of the latter. The type of packing affects the mechanical properties of the metal. (The lines indicating cubic symmetry join spheres on three layers)

(Chapter 1) is an example). This requirement puts a severe restraint on the possible arrangements of the ions and the dense close packing characteristic of metals is never achieved. Largely for that reason, ionic solids are typically less dense than metals. If those two problems are not enough, there are more: many solids have different numbers of cations and anions (calcium chloride, for instance, $CaCl_2$), and in some solids covalent bonding, with its characteristic directional properties, distorts the arrangements expected on the basis of simple sphere-packing. If those four problems, at this stage, are also not enough, then an additional one is that in many cases the ions are far from spherical (sulfate ions, SO_4^{2-}, for instance, are tetrahedral, and there are many more elaborate instances), and all manner of peculiar shapes need to be stacked together in uniform arrays with no overall net electric charge.

Physical chemists have been awed but not daunted by these problems. They have appealed extensively to computers to assemble all manner of crystals from elaborate shapes, building into the software models of how the different components interact with one another and setting the computer to search for the arrangements of lowest energy. This type of study has proved particularly fruitful for the development of the elaborate solids now in vogue as catalysts. These porous solids often have wide channels between blocks of ions to allow the penetration and escape of reactants. The channels effectively increase the surface area enormously by making the reactive hotspots in the interior of the solid accessible to reactant molecules.

Chemical reactivity is not the only aspect of solids in which physical chemists take an interest. Here their interests merge into those of physicists and materials scientists, who will have techniques for computing the electrical, optical, and mechanical properties of even quite complex solids and finding, in collaboration with physical and inorganic chemists (and increasingly with organic chemists), how to enhance the properties in which they are interested. One area that immediately springs to mind where fruitful collaboration lies is the development of the ceramic materials that show the extraordinary property of superconductivity, the ability to conduct electricity without resistance, at temperatures not enormously far below room temperature. Another is the development of electrode and electrolyte materials for the efficient production of electricity by modern batteries.

Intermediate states of matter

I suspect that almost all classifications have frayed edges where an entity exists that is neither fish nor fowl and perhaps a little of both. Frayed frontiers are often regions of considerable interest, where interesting properties lurk and from where new understanding and technologies spring. There are two such frontiers that interest physical chemists, one where liquid meets gas and the other where liquid meets solid.

Liquid meets gas in an interesting way at and around the 'critical point' of a substance. To appreciate the significance of that point, imagine a liquid and its vapour in a sealed, rigid container; we shall imagine it to be transparent so that we can see what is happening. As it is heated, the liquid evaporates and consequently the density of the vapour above it increases. There will come a stage when the density of the vapour matches that of the liquid and the surface between them disappears. This is the critical point. The substance in the vessel is neither liquid nor gas but a *supercritical fluid*. A supercritical fluid has a density comparable to that of a liquid but fills the container it occupies, like a gas.

Supercritical fluids are seriously interesting not only for their intrinsic properties but for their potential as solvents. Among the most heavily studied is supercritical carbon dioxide, $scCO_2$, as a non-noxious, environmentally benign solvent for many organic compounds and which can be removed without leaving any contamination simply by allowing it to vaporize. Supercritical carbon dioxide occurs at around 31°C at pressures of around 73 atm (1 atm is normal atmospheric pressure at sea level).

Supercritical carbon dioxide is at the centre of attention for an increasing number of solvent-based processes. The transport properties of any supercritical fluid (its viscosity and thermal conductivity, for instance) depend strongly on its density, which in turn is sensitive to the pressure and temperature: the density of supercritical carbon dioxide can be adjusted from a gas-like 0.1 g/cm^3 to a liquid-like 1.2 g/cm^3. The solubility of a solute is strongly dependent on the density of the supercritical fluid, so small increases in pressure, particularly close to the critical point, can have very large effects on solubility.

Physical chemists are central to the establishment of the conditions under which a substance can be regarded as supercritical and the exploration of the properties of such fluids and the processes that occur in them is a potentially vigorous area of research. Water itself goes supercritical, but the conditions (around 374°C and 218 atm) are less favourable than for carbon dioxide. Nevertheless, it is a tantalisingly interesting domain for experimental and theoretical exploration.

Physical chemists have become very interested in 'soft matter' (also called 'complex fluids') which lies in character between hard solids and liquids. Many polymeric materials are of this kind, as are surface coatings, adhesives, foams, emulsions, and many biological materials.

Liquid crystals, one form of soft matter, are familiar from electronic displays. They form a so-called *mesophase* state of matter with properties related to those of both liquids and solids. To see liquid crystals in context, recall that in a solid there is long-range order: serried rank follows serried rank right to the edge of a crystal. In contrast, a liquid has only short-range order: there is predictable structure of the first shell of molecules around a central molecule, less predictable in the next shell, and barely any structure for shells of molecules much further away. In a liquid crystal there is long-range order in one or two dimensions but only short-range order in the remaining dimension. For instance, in one type of liquid crystal, the long thin molecules that lie in a plane in a regular pattern, but the parallel planes above and below lie in orientations that show little relation to that central plane (Figure 16).

Liquid crystals have interesting optical properties, as can be inferred from their use in optical displays. Unlike liquids, they also have viscous and mechanical properties that depend on direction. Physical chemists are closely involved in the development of various types of liquid crystal properties and establishing the dynamical properties of the molecules within and between the layers. This type of examination also impinges on biology, for liquid crystals are very similar to the membranes that encapsulate the cells of organisms, including our own.

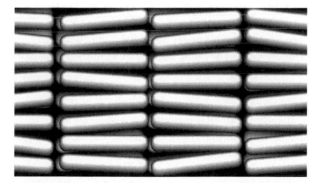

16. In one form of a liquid crystal (the 'nematic' phase) molecules lie in orderly ranks in each plane but neighbouring planes adopt different orientations

One great class of materials that contribute to the class of 'soft matter' substances are the polymers, such as polyethylene, nylon, and all their elaborations. Polymers are a great romping ground for physical chemists as their structures and properties vary so widely. They vary mechanically from the rigid to the runny, electrically from insulator to conductor, and optically too. Some of their properties are modified by the application of electric fields. They represent challenges for their characterization, such as their molecular weights (see the next chapter for a comment), the spread of molecular weights in a sample, the manner in which they might roll up into tight balls or unroll into gangly strands, and the length and conformations of the polymer chains.

The current challenge

Soft polymeric materials are beginning to provide physical chemists with the opportunity to modify materials, including the colours of fabrics, with electric fields, and are offering great opportunities for imaginative investigation both experimental and theoretical. Soft matter presents challenges about structure and its relation to properties, especially the response of samples to sudden changes in the conditions, including mechanical impact. To investigate responses such as these it is essential to build models of the dynamical behaviour of the molecular components and then to convert the theoretical description of that behaviour into a description of the behaviour of bulk samples.

Hard matter also presents challenges, especially when it has fascinating optical and electronic properties, such as superconductivity, and has potential for the storage and manipulation of information. Almost literally at the interface of solids lies graphene, the closest physical chemists come to inhabiting Flatland. Graphite itself is stacks of layers of carbon atoms in a chicken-wire arrangement: these layers slide over each other when impurities are present, which is why impure graphite can be used as a lubricant. Graphene is one of these sheets in isolation. The remarkable strength of this material, and its intriguing electrical properties are both of great interest and are beginning to be studied by physical chemists who now are confronted by the challenges represented by this essentially two-dimensional material. Similarly intriguing materials are aperiodic crystals, which though built from atoms that are packed together closely, do not have the essentially infinite order characteristics of conventional crystals.

Chapter 5
Changing the state of matter

Physical chemists are interested not only in the states of matter themselves but also the transformations they undergo from one form to another, as in the familiar processes of freezing and boiling, which chemists regard as a part of their territory. I shall include dissolving, for observations on the dissolution of gases in liquids were made in the very early days of physical chemistry and are still relevant today, in environmental studies, anaesthesia, respiration, and recreation as well as in the formal structure of physical chemistry.

It should have become apparent from earlier chapters, particularly Chapters 2 and 3, that physical chemistry plays a great deal of attention to processes at equilibrium. In chemistry we are concerned with *dynamic* equilibria in which the forward and reverse processes continue but at matching rates and there is no net change. Chemical equilibria are living equilibria in the sense that because the underlying processes are still active, they are able to respond to changes in the conditions. This aspect of equilibrium is hugely important for both chemical equilibrium (of reactions) and physical equilibrium, the changes of state that I consider here.

Boiling and freezing

Why, a physical chemist might ask, does one form of matter change into another when the conditions are changed? The answer, at one level, must lie in thermodynamics, which identifies the direction of spontaneous change. As we saw in Chapter 2, the signpost of the direction of spontaneous change is the Gibbs energy. In discussions of equilibria, however, physical chemists find it appropriate to use a property introduced by Josiah Gibbs in his original formulation of chemical thermodynamics: the *chemical potential*, μ (mu). As far as we are concerned, the chemical potential can be thought of as the Gibbs energy possessed by a standard-size block of sample. (More precisely, for a pure substance the chemical potential is the molar Gibbs energy, the Gibbs energy per mole of atoms or molecules.) The name is very evocative, for the chemical potential can be thought of as the chemical pushing power of a substance. Thus, if the chemical potential of a liquid is greater than that of a vapour, then the liquid has a tendency to form the vapour. If the opposite is true, then the push is in the opposite direction and the vapour has a tendency to condense into the liquid. If the chemical potentials, the pushing powers, of vapour and liquid are equal, then the two states of matter are in balance: they are in equilibrium and there is no net tendency for either vaporization or condensation. Think tug-of-war, with pushing in place of pulling.

All kinds of changes of state and the corresponding equilibria that are reached in transitions between the states can be expressed in terms of the chemical potential and the pushing power it represents. Physical chemists conjure with expressions for the chemical potential of a substance to identify the conditions in which any two (or more) states of matter are in equilibrium. For instance, they might fix the pressure at 1 atm (normal atmospheric pressure at sea level) and then vary the value of the temperature in their equations until the chemical potentials of the liquid and vapour states coincide: that temperature is the 'boiling point' of the substance. Alternatively, they might vary the temperature in the expressions for the chemical potentials of the liquid and solid forms of the substance and look for the temperature at which the two chemical potentials are equal. At that temperature the solid and liquid forms of the substance are in equilibrium and is called the 'freezing point' of the substance.

Of course, a well-rounded physical chemist will always have in mind what is going on at a molecular level to cause the chemical potentials to change. Chemical potentials, like the Gibbs energy itself, are disguised forms of the total entropy of the system and its surroundings. If the chemical potential of the vapour is lower than that of the liquid (signalling a tendency to vaporize), it really means that the entropy of the universe is greater after vaporization than before it. There are two contributions to that increase. One is the increase in the entropy of the system that accompanies the dispersal of a compact liquid as a gas. The other is the reduction in entropy of the surroundings as heat flows from them into the system to enable the molecules to break free from their neighbours. That reduction in entropy works against the increase in entropy of the system and might well overwhelm it. In that case the total entropy is lower after vaporization and the spontaneous direction of change is in the opposite direction, condensation. However, you should recall from Chapter 2 that a change in entropy is calculated by dividing the heat supplied or lost by the temperature at which the transfer takes place. It follows that if the temperature is raised, then the decrease in entropy of the surroundings will be smaller (the same heat transfer is being divided by a larger number). There will come a point when the temperature is so high that the net change switches from a decrease to an increase (Figure 17). Vaporization then becomes spontaneous.

Physical chemistry has exposed why substances vaporize when the temperature is raised, and the answer is really quite extraordinary:

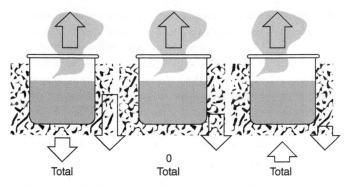

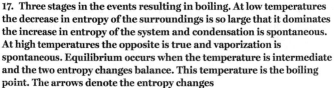

Total   0   Total
              Total

17. **Three stages in the events resulting in boiling. At low temperatures the decrease in entropy of the surroundings is so large that it dominates the increase in entropy of the system and condensation is spontaneous. At high temperatures the opposite is true and vaporization is spontaneous. Equilibrium occurs when the temperature is intermediate and the two entropy changes balance. This temperature is the boiling point. The arrows denote the entropy changes**

the increase in temperature lowers the entropy decrease in the surroundings to the point where it no longer dominates the entropy increase of the system. In effect, raising the temperature soothes changes in the surroundings so that the entropy change in the system becomes dominant. The boiling point, and the freezing point by a similar argument, are manifestations of our ability to manipulate the entropy change in the surroundings by modifying their temperature.

Now consider the role of pressure. One of the earliest successes of chemical thermodynamics, one that suggested to the Victorians that they were on the right track with the emerging science of energy, was the prediction made by Benoit (Emile) Clapeyron (1799–1864) of the effect of pressure on the freezing point of liquids. The intuitive picture is quite clear. For most liquids the sample contracts when it freezes, so (according to Le Chatelier's principle and in practice) increasing the pressure favours the solid and the temperature does not need to be lowered so much for freezing to occur. That is, the application of pressure raises the freezing point. Water, as in most things, is anomalous, and ice is less dense than liquid water, so water expands when it freezes (icebergs float; the *Titanic* was sunk by this anomaly). In this case, the application of pressure favours the liquid and so freezing is achieved only by lowering the temperature even further. That is, for water the application of pressure lowers the freezing point. That response contributes to the advance of glaciers, where the ice melts when pressed against the sharp edges of underlying rocks.

Clapeyron's calculation made predictions about the effect of pressure on the melting point of water in terms of the difference in density of ice and liquid water, and his predictions were not only in the correct direction (pressure lowers the freezing point) but also close to the numerical value observed. Physical chemists now know how to set up the general expression: they can calculate the effect of pressure on the chemical potentials of ice and water and then decide how to adjust the temperature so that the two chemical potentials remain equal when the pressure is changed. They can carry out similar calculations on the effect of pressure on boiling points (increasing the pressure invariably increases the boiling point).

The phase rule

One very special aspect of boiling and freezing is a conclusion from thermodynamics concerning the equilibria between various forms of matter. This is the *phase rule* of Josiah Gibbs, one of the most elegant of all conclusions in chemical thermodynamics. He formulated it in the late 1870s, the golden age of chemical thermodynamics.

A 'phase' is a form of matter, such as a liquid, solid, or gas. It is more specific than 'physical state' because a solid might exist in several different phases. For instance, graphite is one solid phase of carbon, diamond is another. Only helium is known to have two liquid phases, one a conventional liquid and the other a superfluid, which flows without viscosity. No substance has more than one gaseous phase. Each phase of a substance is the most stable under a range of pressures and temperatures. For instance, the common solid phase of water (ordinary ice; there are about a dozen different forms of ice in fact) is its most stable phase at 1 atm and temperatures below 0°C and the gas phase ('water vapour') is the most stable phase at 1 atm and temperatures above 100°C. By 'most stable' we mean that the phase has the lowest chemical potential and all other phases, with higher chemical potentials, have a spontaneous tendency to change into it. A map can therefore be drawn using pressure and temperature as coordinates that shows the regions of pressure and temperature where each phase is the most stable. An analogy is a map of a continent, where each country or state represents the ranges of pressures and temperatures for which the corresponding phase is the most stable (Figure 18). Such a diagram is called a *phase diagram* and is of the greatest importance in materials science, especially when the

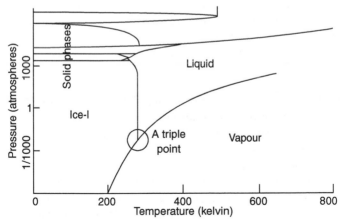

18. A phase diagram, in this case of water, showing the regions of pressure and temperature where each phase is the most stable. There are many forms of ice. Ice-I is the ordinary variety

system under consideration consists of more than one component (iron and several varieties of steel, for instance).

The lines marking the frontiers in a phase diagram are special places, just as they are in actual continents. In a phase diagram they represent the conditions under which the two neighbouring phases are in equilibrium. For example, the line demarking the liquid from the vapour shows the conditions of pressure and temperature at which those two phases are in equilibrium and so it can be regarded as a plot of the boiling temperature against pressure.

There are commonly places in a phase diagram, just as there are in maps of continents, where three phases meet. At this so-called 'triple point' the three phases are in mutual equilibrium, and there are ceaseless interchanges of molecules between all three phases at matching rates. The triple point of a single substance, such as water, is fixed by Nature, and everywhere in the universe (we can suppose) it has exactly the same value. In fact, the triple point of water has been used to define the Kelvin temperature scale, with it (in the current form of the definition) set at 273.16 K exactly; in turn, that definition is used to define the everyday Celsius scale as the temperature on the Kelvin scale minus 273.15 (yes, 273.15, not 273.16).

Gibbs's contribution to understanding phase diagrams was to derive an exceptionally simple rule, the *phase rule*, for accounting for the structure of any phase diagram, not just the simple one-component phase diagram that I have described. The phase rule helps a chemist interpret the diagrams and draw conclusions about the compositions of mixtures, including the changing composition of liquids such as petroleum, in the course of their distillation and purification. Phase diagrams of more elaborate kinds are essential in mineralogy and metallurgy, where they summarize the composition of minerals and alloys.

Dissolving and mixing

The important processes of dissolving and that similar phenomenon, mixing, come within the scope of physical chemistry. Interest in them goes back right to the beginning of physical chemistry and William Henry (1774–1836) who formulated a law of gas solution in 1803.

The mixing of gases is easy to formulate in terms of thermodynamics, and is the starting point for the discussion of mixing and dissolving in general. Physical chemists focus initially on the mixing of perfect gases in which there are no interactions between the molecules: when two gases are allowed to occupy the same container they invariably mix and each spreads uniformly through it. A physical chemist thinks about this mixing as follows.

Because the mixing of two perfect gases is spontaneous, we can infer that the Gibbs energy of any mixture they form must be lower than the total Gibbs energy of the separate gases at the same pressure and temperature, for then mixing is spontaneous in all proportions. There is no change in the entropy of the surroundings: no energy is shipped into or out of the container as heat when the molecules mingle because there are no interactions between them and they are blind to each other's presence. The reason for spontaneous mixing must therefore lie in the increasing entropy of the gases in the container itself. That is perfectly plausible, because the system is less disordered before mixing takes place than after it, when molecules of the two gases are mingled.

Physical chemists import the same reasoning to the class of matter they call an *ideal solution*, the starting point for all discussions of the thermodynamics of mixing of liquids. An ideal solution is very much like a mixture of perfect gases in the sense that it is supposed that mixing occurs without change of energy. Thus, in liquid A the molecules interact with one another and have a certain energy by virtue of these interactions; likewise in liquid B all the molecules interact with one another and also have a certain energy. When the ideal solution forms, A and B molecules are surrounded by one another—around any A there will be A and B molecules in proportions in which the mixture was prepared, and likewise around any B there will be A and B molecules in those proportions—and have a certain energy. In an ideal solution that energy is just the same as in the pure liquids when each molecule is surrounded by its own kind. This approximation, for it is an approximation, implies that when the mixing occurs the molecules mingle together without any change of energy and therefore do not affect the entropy of the surroundings. As in perfect gases, the spontaneity of the mixing is due solely to the increase in entropy of the system as the molecules mingle and the disorder of the system increases.

There aren't many liquid mixtures that behave in this ideal way: the molecules must be very similar for it to be plausible that their interactions are independent of their mixed or unmixed environments: benzene and methylbenzene (toluene) are often cited as plausible examples, but even they are not perfectly ideal. The ideal solution is another example of an idealization in physical chemistry that although a sensible starting point requires elaboration.

The elaboration that has been explored fruitfully in physical chemistry is the *regular solution*. In a regular solution it is supposed that the two types of molecule are distributed through the mixture perfectly randomly, just like in an ideal solution. However, unlike in an ideal solution, the energy of an A molecule does depend on the proportion of A and B molecules that surround it, and likewise for B. This model captures quite a lot of the properties of real solutions. For instance, if the strengths of the A–A and B–B interactions outweigh the A–B interactions, then the liquids will not mix fully and the system consists of two phases, one of A dissolved in an excess of B and the other of B dissolved in an excess of A.

I have spoken of gases dissolved in (mixed with) one another and of liquids dissolved in (mixed with) one another. What about gases dissolved in liquids? This is where Henry made his contribution more than two centuries ago and established a principle that is still used today. Henry found that the quantity of gas that dissolves in any liquid is proportional to the pressure of the gas. That might, to us, seem a rather trivial conclusion from a lifetime's work, but it is worth reflecting on how a physical chemist might view the conclusion in terms of the molecular processes occurring at the interface between the liquid and the gas.

As we saw in Chapter 4, according to the kinetic model, a gas is a maelstrom of molecules in ceaseless motion, colliding with each other billions of times a second. The molecules of a gas above the surface of a liquid are ceaselessly pounding on the surface and, if splashing is appropriate on this scale, splashing down into the liquid. At the same time, molecules of gas already embedded in the liquid are rising to the surface by the ceaseless molecular jiggling in the liquid, and once there can fly off and join their colleagues above the surface (Figure 19). At equilibrium, the rate at which the gas molecules escape from the liquid matches the rate at which molecules splash down into it. At this equilibrium we can report a certain solubility of the gas (such as how many dissolved molecules are present in each volume of liquid). If the gas pressure is increased, the rain of molecules from the gas increases in proportion, but the rate of escape of dissolved molecules remains unchanged. When equilibrium is renewed, more molecules will be found in the liquid, just as Henry's law describes.

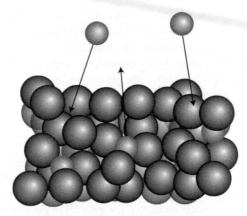

19. A simple explanation of Henry's law is that at equilibrium the rate at which gas molecules strike the surface of a liquid and penetrate into its bulk matches the rate at which those already present leave. The rate at which gas molecules strike the surface of a liquid and penetrate into its bulk, but not the rate at which those already present leave, is increased by raising the pressure of the gas

When the temperature of the liquid is raised, it is easier for a dissolved molecule to gather sufficient energy to escape back up into the gas; the rate of impacts from the gas is largely unchanged. The outcome is a lowering of the concentration of dissolved gas at equilibrium. Thus, gases appear to be less soluble in hot water than in cold. When the eyes of physical chemists waking up in the morning see bubbles in the water on their bedside tables as it has warmed over night, they are, or more realistically should be, reminded of William Henry and his discussion of the dissolving of gases.

Henry's law contributes to our understanding of the network of processes that underlie life. The ability of aquatic life to thrive depends on the presence of dissolved oxygen: the pressure of oxygen in air is sufficient to maintain its concentration at a viable level. However, if the water is warmed by industrial or natural causes, the oxygen level might fall to fatal values. Henry's law also plays a role in recreation, as in scuba diving, and its commercial counterpart deep-sea diving, where the dissolution of oxygen and nitrogen in blood and their possible formation of dangerous bubbles giving rise to 'the bends', can be expressed in its terms.

Transitions of solutions

Physical chemists have worked out how the presence of dissolved substances affects the properties of solutions. For instance, the everyday experience of spreading salt on roads to hinder the formation of ice makes use of the lowering of freezing point of water when a salt is present. There is also another hugely important property of solutions that pure liquids do not exhibit, namely *osmosis*. Osmosis (the word comes from the Greek word for 'push') is the tendency of a solvent to push its way into a solution through a membrane. It is responsible for a variety of biological phenomena, such as the rise of sap in trees and the maintenance of the shape of our red blood cells.

When physical chemists consider the thermodynamics of solutions they have in mind the roles of energy and entropy. I have explained that for vaporization to become spontaneous, the temperature of the surroundings must be increased to reduce the change in their entropy when energy flows out of them as heat and into the liquid. When a substance is dissolved in the liquid, the entropy is greater than when the solvent is pure because it is no longer possible to be sure that a blind selection of a molecule is that of a solute or a solvent: there is more disorder and therefore higher entropy. Because the entropy of the liquid is already higher than before, and the vapour (of the solvent alone, because the solute does not vaporize) has the same entropy, the increase in the entropy of the system when the liquid vaporizes is not as great as for the pure liquid. As usual, the entropy of the surroundings falls because heat flows into the liquid and at low temperatures is so big that it dominates the increase in entropy of the system. To reduce the entropy change of the surroundings to match the now smaller entropy increase of the system, the temperature must be raised, but it must be raised further than before. That is, the boiling point is raised by the presence of a dissolved substance (Figure 20).

Similar reasoning applies to the freezing point, which is lowered by the presence of a solute. It is often said that the addition of antifreeze to a car engine is an example of this lowering of freezing point. Although there are similarities, the function of the antifreeze at the high concentrations used is quite different: the antifreeze molecules simply mingle with the water molecules and prevent their forming bonds and freezing to a solid.

When a liquid and its vapour are present in a closed container the vapour exerts a characteristic pressure (when the escape of molecules from the liquid matches the rate at which they splash back down into it, as in Figure 19). This characteristic pressure depends on the temperature and is called the 'vapour pressure' of the liquid. When a solute is present, the vapour pressure at a given temperature is lower than that of the pure liquid for reasons

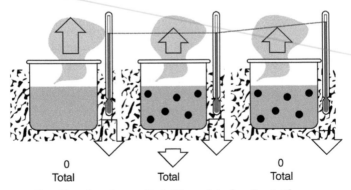

0 Total   Total   0 Total

20. The effect of a solute on the boiling point of a solvent. The temperature must be raised for the entropy decrease of the surroundings to match the now smaller increase in entropy of the solvent when it vaporizes from the solution

related to entropy, more or less as I explained above. The extent of lowering is summarized by yet another limiting law of physical chemistry, this one having been formulated by François-Marie Raoult (1830–1901), who spent much of the later decades of his life measuring vapour pressures. In essence, 'Raoult's law' states that the vapour pressure of a solvent or of a component of a liquid mixture is proportional to the proportion of solvent or liquid molecules present. Mixtures of liquids that obey this law strictly are the 'ideal solutions' that I have already mentioned. Actual solutions, 'real solutions', obey it only in a limiting sense when the concentration of the solute or of the second liquid component goes to zero. As for most limiting laws, Raoult's law is used as the starting point for the discussion of the thermodynamic properties of mixtures and solutions and is the foundation for more elaborate treatments, such as treating the solution as regular rather than ideal.

Osmosis, the tendency of solvent molecules to flow from the pure solvent to a solution separated from it by a permeable membrane (technically, a 'semipermeable membrane', one that allows the transit of solvent molecules but not solute molecules or ions), is another manifestation of the effects of entropy caused by the presence of a solute. The entropy when a solute is present in a solvent is higher than when the solute is absent, so an increase in entropy, and therefore a spontaneous process, is achieved when solvent flows through the membrane from the pure liquid into the solution. The tendency for this flow to occur can be overcome by applying pressure to the solution, and the minimum pressure needed to overcome the tendency to flow is called the 'osmotic pressure'.

If one solution is put into contact with another through a semipermeable membrane, then there will be no net flow if they exert the same osmotic pressures and are 'isotonic'. On the other hand, if a greater pressure than the osmotic pressure is applied to the solution in contact through a semipermeable membrane with the pure solvent, then the solvent will have a tendency to flow in the opposite direction, flowing from solution to pure solvent. This effect is the 'reverse osmosis' that is used to purify sea water and render it potable.

If the solution is ideal, then there is a very simple relation between the osmotic pressure and the concentration of solute, which is summarized by an equation proposed by Jacobus van 't Hoff (1852–1911), the winner of the first Nobel Prize for chemistry (in 1901). His equation, which sets the osmotic pressure proportional to the concentration of solute and the temperature, is yet another of physical chemistry's limiting laws, for it is strictly valid only in the limit of zero concentration of solute. Nevertheless, it is a very useful starting point for the discussion of actual solutions.

The principal application of van 't Hoff's equation and its elaborations is to the determination of the molecular weights of polymers. The problem with such molecules is that they are so huge that even a substantial mass of them does not amount to a concentrated solution (in terms of numbers present). Osmotic pressure, however, is very sensitive to concentration, and their molecular weights can be inferred from its value. One remaining problem, however, is that because they are huge, they form far from ideal solutions, so elaborations of van 't Hoff's equation are essential for the analysis of the results.

Other transitions

More subtle than either boiling or freezing are the transitions between varieties of solid phases, all of which are open to examination by physical chemists, although in many cases they lie more naturally in physics' grasp. Transitions between the various phases of solids include those between different kinds of magnetic behaviour and from metallic to superconducting states. The latter have become highly important since the discovery of the ceramic materials with transition temperatures not very far below room temperature. Solid-to-solid phase transitions are also important in inorganic chemistry, geochemistry, metallurgy, and materials science in general and all these fields collaborate with physical chemists to elucidate the processes involved and to seek ways of expressing the transitions quantitatively.

The current challenge

There are two very subtle kinds of phase transition that are now becoming of great interest to physical chemists. One is the conversion of one kind of matter into another by the process of 'self assembly'. In this process, the intrinsic structure of individual molecules encourages them to assemble into complex structures without external intervention. Even more subtle is the modification of structures locally, which underlies the storage of information. An extreme example is the storage of vast amounts of information, such as the works of Shakespeare, by the synthesis of strands of DNA with an appropriate sequence of bases.

Chapter 6
Changing the identity of matter

A great deal of chemistry is concerned with changing the identity of matter by the deployment of chemical reactions. It should hardly be surprising that physical chemistry is closely interested in the processes involved and that it has made innumerable contributions to our understanding of what goes on when atoms exchange partners and form new substances. Physical chemists are interested in a variety of aspects of chemical reactions, including the rates at which they take place and the details of the steps involved during the transformation. The general field of study of the rates of reactions is called *chemical kinetics*. When the physical chemist gets down to studying the intimate details of the changes taking place between atoms, chemical kinetics shades into *chemical dynamics*.

Chemical kinetics is a hugely important aspect of physical chemistry for it plays a role in many related disciplines. For instance, it is important when designing a chemical plant to know the rates at which intermediate and final products are formed and how those rates depend on the conditions, such as the temperature and the presence of catalysts. A human body is a network of myriad chemical reactions that are maintained in the subtle balance we call homeostasis, and if a reaction rate runs wild or slows unacceptably, then disease and death may ensue. The observation of reaction rates and their dependence on concentration of reactants and the temperature gives valuable insight into the steps by which the reaction takes place, including the role of a catalyst, and how the rates may be optimized. Chemical dynamics deepens this understanding by putting individual molecular changes under its scrutiny.

Then there are different types of chemical reactions other than those achieved simply by mixing and heating. Some chemical reactions are stimulated by light, and physical chemists play a central role in the elucidation of these 'photochemical reactions' in the field they call *photochemistry*. There is no more important photochemical reaction than the photosynthesis that captures the energy of the Sun and stands at the head of the food chain, and physical chemists play a vital role in understanding its mechanism and devising ways to emulate it synthetically. Then there is the vital collaboration of electricity and chemistry in the form of *electrochemistry*. The development of this aspect of chemistry is central to modern technology and the deployment of electrical power.

Spontaneous reaction

The first contribution to understanding chemical reactions that physical chemists make is to identify, or at least understand, the spontaneous direction of reaction. (Recall from Chapter 2 that 'spontaneous' has nothing to do with rate.) That is, we need to know where we are going before we start worrying about how fast we shall get there. This is the role of chemical thermodynamics, which also provides a way to predict the equilibrium composition of the reaction mixture, the stage at which the reaction seems to come to a halt with no tendency to change in either direction, towards more products or back to reactants. Remember, however, that chemical equilibria are dynamic: although change appears to have come to an end, the forward and reverse reactions are continuing, but now at matching rates. Like the equilibria we encountered in Chapter 5, chemical equilibria are living, responsive conditions.

The natural direction of a chemical reaction corresponds to time's usual arrow: the direction of increasing entropy of the universe. As explained in Chapter 2, provided the pressure and temperature are held constant, attention can be redirected from the entire universe to the flask or test-tube where their reaction is taking place by focusing instead on the Gibbs energy of the reaction system. If at a certain stage of the reaction the Gibbs energy decreases as products are formed, then that direction is spontaneous and the reaction has a tendency to go on generating products. If the Gibbs energy increases, then the reverse reaction is spontaneous and any products present will have a tendency to fall apart and recreate the reactants. If the Gibbs energy does not change in either direction, then the reaction is at equilibrium.

These features can all be expressed in terms of the chemical potential of each participant of the reaction, the concept introduced in Chapter 5 in connection with physical change. The name 'chemical potential' now fulfils its significance. Thus, the chemical potentials of the reactants effectively push the reaction towards products and those of the products push it back towards reactants. The reaction is like a chemical tug-of-war with the chemical potentials the opponents pushing rather than pulling: stalemate is equilibrium. Physical chemists know how the chemical potentials change as the composition of the reaction mixture changes, and can use that information to calculate the composition of the mixture needed for the chemical potentials of all the participants to be equal and the reaction therefore to be at equilibrium. I have already explained (in Chapter 3) that the mixing of the reactants and products plays a crucial role in deciding where that equilibrium lies.

Physical chemists don't deduce the equilibrium composition directly. The thermodynamic expressions they derive are in terms of a combination of the concentrations of the reactants and products called the *reaction quotient*, Q. Broadly speaking, the reaction quotient is the ratio of concentrations, with product concentrations divided by reactant concentrations. It takes into account how the mingling of the reactants and products affects the total Gibbs energy of the mixture. The value of Q that corresponds to the minimum in the Gibbs energy can be found, and then the reaction tug-of-war is in balance and the reaction is at equilibrium. This particular value of Q is called the *equilibrium constant* and denoted K. The equilibrium constant, which is characteristic of a given reaction and depends on the temperature, is central to many discussions in chemistry. When K is large (1000, say), we can be reasonably confident that the equilibrium mixture will be rich in products; if K is small (0.001, say), then there will be hardly any products present at equilibrium and we should perhaps look for another way of making them. If K is close to 1, then both reactants and products will be abundant at equilibrium and will need to be separated.

The equilibrium constant is, as this discussion implies, related to the change in Gibbs energy that accompanies the reaction, and the relation between them is perhaps one of the most important in chemical thermodynamics. It is the principal link between measurements made with calorimeters (which, as explained in Chapter 2, are used to calculate Gibbs energies) and practical chemistry, through a chemist's appreciation of the significance of equilibrium constants for understanding the compositions of reaction mixtures.

Comment

This important equation is $\Delta_r G^\ominus = -RT \ln K$, where $\Delta_r G^\ominus$ is the change in Gibbs energy on going from pure reactants to pure products and T is the temperature; R is the gas constant, $R = N_A k$.

Equilibrium constants vary with temperature but not, perhaps surprisingly, with pressure. The hint that they depend on temperature comes from remembering that the Gibbs energy includes the temperature in its definition ($G = H - TS$). Jacobus van't Hoff, who was mentioned in Chapter 5 in relation to osmosis, discovered a relation between the sensitivity of the equilibrium constant to temperature and the change in enthalpy that accompanies the reaction. In accord with Le Chatelier's principle about the self-correcting tendency of equilibria exposed to disturbances, van't Hoff's equation implies that if the reaction is strongly exothermic (releases a lot of energy as heat when it takes place), then the equilibrium constant decreases sharply as the temperature is raised. The opposite is true if the reaction is strongly endothermic (absorbs a lot of energy as heat). That response lay at the root of Haber and Bosch's scheme to synthesize ammonia, for the reaction between nitrogen and hydrogen is strongly exothermic, and although they needed to work at high temperatures for the reaction to be fast enough to be economical, the equilibrium constant decreased and the yield would be expected to be low. How they overcame this problem I explain below.

Reaction rate

As I have emphasized, thermodynamics is silent on the rates of processes, including chemical reactions, and although a reaction might be spontaneous, that tendency might be realized so slowly in practice that for all intents and purposes the reaction 'does not go'. The role of chemical kinetics is to complement the silence of thermodynamics by providing information about the rates of reactions and perhaps suggesting ways in which rates can be improved.

What is meant by the 'rate' of a chemical reaction? Rate is reported by noting the change in concentration of a selected component and dividing it by the time that change took to occur. Just as the speed of a car will change in the course of a journey, so in general does the rate of a reaction. Therefore, to get the 'instantaneous rate' of a reaction (like the actual speed of a car at any moment), the time interval between the two measurements of concentration needs to be very small. There are technical ways of implementing this procedure that I shall not dwell on. Typically it is found that the rate of a reaction decreases as it approaches equilibrium.

That the instantaneous rate (from now on, just 'rate') depends on concentration leads to an important conclusion. It is found experimentally that many reactions depend in a reasonably simple way on the concentrations of the reactants and products. The expression describing the relation between the rate and these concentrations is called the *rate law* of the reaction. The rate law depends on one or more parameters, which are called *rate constants*. These parameters are constant only in the sense that they are independent of the concentrations of the reactants and products: they do depend on the temperature and typically increase with increasing temperature to reflect the fact that most reactions go faster as the temperature is increased.

Some rate laws are very simple. Thus it is found that the rates of some reactions are proportional to the concentration of the reactants and others are proportional to the square of that concentration. The former are classified as 'first-order reactions' and the latter as 'second-order reactions'. The advantage of this classification, like all classifications, is that common features can be identified and applied to reactions in the same class. In the present case, each type of reaction shows a characteristic variation of composition with time. The concentration of the reactant in a first-order reaction decays towards zero at a rate determined by its rate constant; that of a second-order reactions decays to zero too, but although it might start of at the same rate, the concentration of reactant takes much longer to reach zero. An incidental point in this connection is that many pollutants in the environment disappear by a second-order reaction, which is why low concentrations of them commonly persist for long periods.

One of the points to note is that it isn't possible to predict the order of the reaction just by looking at its chemical equation. Some very simple reactions have very complicated rate laws; some that might be expected to be first order (because they look as though a molecule is just shaking itself apart) turn out to be second order, and some that might be suspected to be second order (for example, they look as though they take place by collisions of pairs of molecules) turn out to be first order.

One aim of chemical kinetics is to try to sort out this mess. To do so, a *reaction mechanism* (a sequence of steps involving individual molecules) is proposed that is intended to represent what is actually happening at a molecular level. Thus, it might be supposed that in one step two molecules collide and that one runs off with much more energy than it had before, leaving the other molecule depleted in energy. Then in the second step that excited molecule shakes itself into the arrangement of atoms corresponding to the product. The rate laws for these individual steps (and these laws can be written down very simply: in this case the first step is second order and the second step is first order) are combined to arrive at the rate law for the overall reaction. If the rate law matches what is observed experimentally, then the proposed mechanism is plausible; if it doesn't match, then a different mechanism has to be proposed. The problem is that even if the overall rate law is correct, there might be other mechanisms that lead to the same rate law, so other evidence has to be brought to bear to confirm that the mechanism is indeed correct. In that respect, the establishment of a reaction mechanism is akin to the construction of proof of guilt in a court of law.

In some cases it is possible to identify a step that controls the rate of the entire overall process. Such a step is called the *rate-determining step*. Many steps might precede this step, but they are fast and do not have a significant influence on the overall rate. A mechanism with a rate-determining step has been likened to having six-lane highways linked by a one-lane bridge.

Reaction rate and temperature

Most reactions go faster when the temperature is raised. The reason must be found in the rate constants in the rate law, for only they depend on temperature. The specific temperature dependence that is typical of rate constants was identified by Svante Arrhenius (1859–1927) who proposed in 1889 that their dependence on temperature could be summarized by introducing two parameters, the more interesting of which became known as the *activation energy* of the reaction: reactions with high activation energies proceed slowly at low temperatures but respond sharply to changes of temperature.

Comment

The Arrhenius expression is $k_r = Ae^{-E_a/RT}$, where k_r is the rate constant and E_a is the activation energy; A is the second parameter.

The simplest explanation of the so-called 'Arrhenius parameters' is found in the *collision theory* of gas-phase reactions. In this theory, the rate of reaction depends on the rate at which molecules collide: that rate can be calculated from the kinetic theory of gases, as explained in Chapter 4. However, not all the collisions occur with such force that the atoms in the colliding molecules can rearrange into products. The proportion of high-energy collisions can be calculated from the Boltzmann distribution (Chapter 3). Then, the expression developed by Arrhenius is obtained if it is supposed that reaction occurs only if the energy of the collision is at least equal to the activation energy (Figure 21). Thus, the activation energy is now interpreted as the minimum energy required for reaction. Increasing the temperature increases the proportion of collisions that have at least this energy, and accordingly the reaction rate increases.

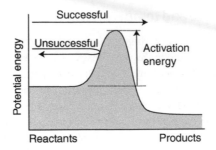

21. The activation energy is the minimum energy required for reaction and can be interpreted as the height of a barrier between the reactants and products: only molecules with at least the activation energy can pass over it

Gas-phase reactions are important (for instance, for accounting for the composition of atmospheres as well as in a number of industrial processes), but most chemical reactions take place in solution. The picture of molecules hurtling through space and colliding is not relevant to reactions in solution, so physical chemists have had to devise an alternative picture, but one that still accounts for the Arrhenius parameters and particularly the role of the activation energy.

The picture that physical chemists have of reactions in solution involves two processes. In one, the reactants jostle at random in the solution and in due course their paths may bring them into contact. They go on jostling, and dance around each other before their locations diverge again and they go off to different destinies. The second process occurs while they are dancing around each other. It may be that the ceaseless buffeting from the solvent molecules transfers to one or other of the reactant molecules sufficient energy—at least the activation energy—for them to react before the dance ends and they separate.

In a 'diffusion-controlled reaction' the rate-determining step is the jostling together, the diffusion through the solution, of the reactants. The activation energy for the reaction step itself is so low that reaction takes place as soon as the reactants meet. In this case the metaphorical narrow bridge precedes the six-lane highway, and once the reactants are over it the journey is fast. The diffusion together is the rate-determining step. Diffusion, which involves molecules squirming past one another, also has a small activation energy, so Arrhenius-like behaviour is predicted. In an 'activation-controlled reaction' the opposite is true: encounters are frequent but most are sterile because the actual reaction step requires a lot of energy: its activation energy is high and constitutes the rate-determining step. In this case, the six-lane highway leads up to the narrow bridge.

It is much harder to build a model of the reaction step in solution than it is to build the model for gas-phase reactions. One approach is due to Henry Eyring (1901–61), who provides another scandalous example of a Nobel Prize deserved but not awarded. In his *transition-state theory*, it is proposed that the reactants may form a cluster of atoms before falling apart as the products. The model is formulated in terms of statistical thermodynamic quantities, with the ubiquitous Boltzmann distribution appearing yet again. Transition-state theory has been applied to a wide variety of chemical reactions, including those responsible for oxidation and reduction and electrochemistry. The principal challenge is to build a plausible model of the intermediate cluster and to treat that cluster quantitatively.

Catalysis

A *catalyst* is a substance that facilitates a reaction. The Chinese characters for it make up 'marriage broker', which conveys the sense very well. Physical chemists are deeply involved in understanding how catalysts work and developing new ones. This work is of vital economic significance, for just about the whole of chemical industry throughout the world relies on the efficacy of

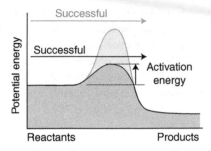

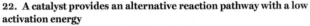

22. A catalyst provides an alternative reaction pathway with a low activation energy

catalysts. Living organisms owe a similar debt, for the protein molecules we call enzymes are highly effective and highly selective catalysts that control almost every one of the myriad reactions going on inside them and result in them 'being alive'.

All catalysts work by providing a different, faster pathway for a reaction. In effect, they provide a pathway with an activation energy that is lower than in the absence of the catalyst (Figure 22). Most industrial catalysts work by providing a surface to which reactant molecules can attach and in the process of attachment perhaps be torn apart. The resulting fragments are ripe for reaction, and go on to form products. For physical chemists to contribute to the development of catalysts, they need to understand the structures of surfaces, how molecules attach to (technically: 'adsorb on') them, and how the resulting reactions proceed. They employ a variety of techniques, with computation used to model the reaction events and observation used to gather data. A surge in their ability to examine surfaces took place a few years ago with the development of the techniques I describe in Chapter 7 that provide detailed images of surfaces on an atomic scale.

Biological catalysts, enzymes, require different modes of investigation, but with computation still a major player and physical chemists in collaboration with molecular biologists, biochemists, and pharmacologists. Nowhere in chemistry is shape more central to function than in the action of enzymes, for instead of catalysis occurring on a reasonably flat and largely featureless surface, an enzyme is a highly complex structure that recognizes its prey by shape, accepts it into an active region of the molecule, produces change, and then delivers the product on to the next enzyme in line. Disease occurs when an enzyme ceases to act correctly because its shape is modified in some way and becomes unable to recognize its normal prey and perhaps acts on the wrong molecule or when its active site is blocked by an intrusive molecule and it fails to act at all. The recognition of the important role of shape provides a hint about how a disease may be cured, for it may be possible to build a molecule that can block the action of an enzyme that has gone wild or reawaken a dormant enzyme. Physical chemists in collaboration with pharmacologists make heavy users of computers to model the docking of candidate molecules into errant active sites.

Photochemistry

Photochemistry plays an important role in physical chemistry not only because it is so important (as in photosynthesis) but also because the use of light to stimulate a reaction provides very precise control over the excitation of the reactant molecules. The field has been greatly influenced by the use of lasers with their sharply defined frequencies and their ability to generate very short pulses of radiation. The latter enables the progress of reactions to be followed on very short timescales, with current techniques easily dealing with femtosecond phenomena (1 fs = 10^{-15} s) and progress being made with attosecond phenomena too (1 as = 10^{-18} s), where chemistry is essentially frozen into physics.

The idea behind photochemistry is to send a short, sharp burst of light into a sample and to watch what happens. There are physical and chemical consequences.

The physical consequences are the emission of light either as *fluorescence* or as *phosphorescence*. The two forms of emission are distinguished by noting whether the emission persists after the illuminating source is removed: if emission ceases almost immediately it is classified as fluorescence; if it persists, then it is phosphorescence. Physical chemists understand the reasons for the difference in this behaviour, tracing it to the way that the energetically excited molecules undergo changes of state. In fluorescence, the excited state simply collapses back into the ground state. In phosphorescence, the excited molecule switches into another excited state that acts like a slowly leaking reservoir.

Fluorescence is eliminated if another molecule diffuses up to the excited molecule and kidnaps its excess energy. This 'quenching' phenomenon is used to study the motion of molecules in solution, for by changing the concentration of the quenching molecule and watching its impact on the intensity of the fluorescence inferences can be made about what is taking place in the solution. Moreover, the decay of the fluorescence after the stimulating illumination is extinguished is not instantaneous, and the rate of decay, though rapid, can be monitored with or without quenching molecules present and information about the rates of the various processes taking place thereby deduced.

The chemical consequences of photoexcitation stem from the fact that the excited molecule has so much energy that it can participate in changes that might be denied to its unexcited state. Thus, it might be so energy-rich that it shakes itself apart into fragments. These fragments might be 'radicals' which are molecules with an unpaired electron. Recall from Chapter 1 that a covalent bond consists of a shared pair of electrons, which we could denote A:B. If a molecule falls apart into fragments, one possibility is that the electron pair is torn apart and two radicals, in this case A· and ·B, are formed. Most radicals are virulent little hornets, and attack other molecules in an attempt to re-form an electron pair. That attack, if successful, leads to the formation of more radicals, and the reaction is propagated, sometimes explosively. For instance, hydrogen and chlorine gases can be mixed and cohabit in a container without undergoing change, but a flash of light can stimulate an explosive reaction between them.

In some cases the photoexcited molecule is just an intermediary in the sense that it does not react itself but on collision with another molecule passes on its energy; the recipient molecule then reacts. Processes like this, including the radical reactions, are enormously important for understanding the chemical composition of the upper atmosphere where molecules are exposed to energetic ultraviolet radiation from the Sun. The reactions are built into models of climate change and the role of pollutants in affecting the composition of the atmosphere, such as the depletion of ozone (O_3).

Photochemical processes can be benign. There is no more benign photochemical process than photosynthesis, in which photochemical processes capture the energy supplied by the distant Sun and use it to drive the construction of carbohydrates, thus enabling life on Earth. Physical chemists are deeply involved in understanding the extraordinary role of chlorophyll and other molecules in this process and hoping to emulate it synthetically in what would be a major contribution to resolving our energy problems on Earth and wherever else in due course we travel. Meanwhile, physical chemists also contribute to understanding the photovoltaic materials that are a temporary but important approximation to the power of photosynthesis and make use of the excitation of electrons in inorganic materials.

Electrochemistry

Another major contribution that physical chemistry makes to the survival of human societies and technological advance is electrochemistry, the use of chemical reactions to generate electricity (and the opposite process, electrolysis). A major class of chemical reactions of interest in electrochemistry consists of *redox reactions*, and I need to say a word or two about them.

In the early days of chemistry, 'oxidation' was simply reaction with oxygen, as in a combustion reaction. Chemists noticed many similarities to other reactions in which oxygen was not involved, and came to realize that the common feature was the removal of electrons from a substance. Because electrons are the glue that holds molecules together, in many cases the removal of electrons resulted in the removal of some atoms too, but the core feature of oxidation was recognized as electron loss.

In similar former days, 'reduction' was the term applied to the extraction of a metal from its ore, perhaps by reaction with hydrogen or (on an industrial scale in a blast furnace) with carbon. As for oxidation, it came to be recognized that there were common features in other reactions in which metal ores were not involved and chemists identified the common process as the addition of electrons.

Oxidation is now defined as electron loss and reduction is defined as electron gain. Because electron loss from one species must be accompanied by electron gain by another, oxidation is always accompanied by reduction, and their combinations, which are called 'redox reactions', are now recognized as the outcome of electron transfer, the transfer perhaps being accompanied by atoms that are dragged along by the migrating electrons.

The core idea of using a chemical reaction to produce electricity is to separate reduction from oxidation so that the electrons released in the oxidation step have to migrate to the location where reduction takes place. That migration of electrons takes place through a wire in an external circuit and constitutes an electric current driven by the spontaneous reaction taking place in the system (Figure 23). All batteries operate in this way, with the electrodes representing the gateways through which the electrons

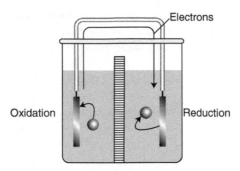

Electrons

Oxidation

Reduction

23. In an electrochemical cell, the sites of oxidation (electron loss) and reduction (electron gain) are separated, and as the spontaneous reaction proceeds electrons travel from one electrode to the other

enter or leave each cell. Electrolysis is just the opposite: an external source pulls electrons out of the cell through one electrode and pushes them back into the cell through another, so forcing a non-spontaneous redox reaction to occur.

There are 'academic' and 'industrial' applications of electrochemistry. I pointed out in Chapter 2 that the change in Gibbs energy during a reaction is an indication of the maximum amount of non-expansion work that the reaction can do. Pushing electrons through a circuit is an example of non-expansion work, so the potential difference that a cell can generate (its 'voltage'), a measure of the work that the cell can do, is a way of determining the change in Gibbs energy of a reaction. You can now see how a thermodynamicist views a 'flat' battery: it is one in which the chemical reaction within it has reached equilibrium and there is no further change in Gibbs energy. Physical chemists know how the Gibbs energy is related to the concentrations of the reactants and products (through the reaction quotient, Q), and can adapt this information to predict the voltage of the cell from its composition. If you recall from Chapter 2, I remarked that thermodynamics provides relations, sometimes unexpected ones, between disparate properties. Electrochemists make use of this capability and from how the voltage changes with temperature are able to deduce the entropy and enthalpy changes due to the reaction, so electrochemistry contributes to thermochemistry.

Electrochemists are also interested in the rate at which electrons are transferred in redox reactions. An immediate application of understanding the rates at which electrons are transferred into or from electrodes is to the improvement of the power, the rate of generation of electrical energy, that a cell can produce. But the rates of electron transfer also have important biological implications, because many of the processes going on inside organisms involve the transfer of electrons. Thus, if our brains are operating at 100 watts and there are potential differences within us amounting to about 10 volts, then we are 10 ampere creatures. The understanding of the rate of electron transfer between biologically important molecules is crucial to an understanding of the reactions, for instance, that constitute the respiratory chain, the deployment of the oxygen that we breathe and which drives all the processes within us.

The industrial applications of electrochemistry are hugely important, for they include the development of light, portable, powerful batteries and the development of 'fuel cells'. A fuel cell is simply a battery in which the reactants are fed in continuously (like a fuel) from outside, rather than being sealed in once and for all at manufacture. Physical chemists contribute to the improvement of their efficiencies by developing new electrode and electrolyte materials (the electrolyte is the medium where the redox reactions take place). Here lie opportunities for the union of electrochemistry and photochemistry, for some cells being developed generate electricity under the influence of sunlight rather than by using a physical fuel.

The metal artefacts fabricated at such expense (of cash and energy) crumble in the face of Nature and the workings of the Second Law. Corrosion, the electrochemical disease of metal, destroys and needs to be thwarted. Electrochemists seek an understanding of corrosion, for it is a redox reaction and within their domain, and through that understanding look for ways to mitigate it and thereby save societies huge sums.

Chemical dynamics

Chemical dynamics is concerned with molecular intimacies, where physical chemists intrude into the private lives of atoms and watch in detail the atomic processes by which one substance changes into another.

One of the most important techniques for discovering what is happening to individual molecules when they react is the *molecular beam*. As its name suggests, a molecular beam is a beam of molecules travelling through space. It is just not any old beam: the speeds of the molecules can be selected in order that the kinetic energy they bring to any collision can be monitored and controlled. It is also possible to select their vibrational and rotational states and to twist them into a known orientation. In a molecular beam apparatus, one beam impinges on target molecules, which might be a puff of gas or might be another beam. The incoming beam is scattered by the collision or, if the impacts are powerful enough, its molecules undergo reactions and the products fly off in various directions. These products are detected and their states determined. Thus, a very complete picture can be constructed of the way that molecules in well-defined states are converted into products in similarly well-defined states.

Molecular dynamics is greatly strengthened by its alliance with computational chemistry. As two molecules approach each other, their energy changes as bonds stretch and break, they bend, and new bonds are formed. The calculation of these energies is computationally very demanding but has been achieved in a number of simple cases. It results in a landscape of energies known as a *potential-energy surface*. When molecules travel through space they are, in effect, travelling across this landscape. They adopt paths that can be calculated either by using Newton's laws of motion or, in more sophisticated and appropriate treatments, by using the laws of quantum mechanics (that is, by solving Schrödinger's equation).

Close analysis of the paths that the molecules take through the landscape gives considerable insight into the details of what takes place at an atomic level when a chemical reaction occurs. Thus it may be found that reaction is more likely to ensue if one of the molecules is already vibrating as it approaches its target rather than just blundering into it with a lot of translational energy. Reaction might also be faster if a molecule approaches the target from a particular direction, effectively finding a breach in its defensive walls.

Physical chemists need to forge a link between this very detailed information and what happens on the scale of bulk matter. That is, they need to be able to convert the trajectories that they compute across the potential-energy landscape into the numerical value of a rate constant and its dependence on temperature. This bridge between the microscopic and the macroscopic can be constructed, and there is increasing traffic across it.

The current challenge

The challenge that confronts the physical chemists who experiment with molecular beams and their intellectual compatriots who compute the potential energy surfaces on which the interpretation of the reactive and non-reactive scattering depend is to find ways to transfer their techniques and conclusions into liquids, the principal environment of chemical reactions. Although there is little hope of doing so directly (although the continuing massive advances in computing power rule nothing out), the detailed knowledge obtained from the gas phase adds to our insight about what is probably happening in solution and generally enriches our understanding of chemistry.

At an experimental level perhaps nothing is more important to the development of energy-hungry civilization than the role that physical chemistry plays in the development of the generation and deployment of electric power through photochemistry and the improvement of fuel cells and portable sources. A major part of this application is the development of an understanding of surface phenomena, seeking better catalysts, developing better electrode materials, and exploring how photochemical activity can be enhanced. Nature has much to teach in this connection, with clues obtained from the study of photosynthesis and the action of enzymes.

Molecular beams are capable of providing exquisite information about the details of reaction events and laser technology, now capable of producing extraordinarily brief flashes of radiation, permits reaction events to be frozen in time and understood in great detail.

Classical kinetic techniques continue to be important: the observation of reaction rates continues to illuminate the manner in which enzymes act and how they may be inhibited. Some reactions generate striking spatial and sometimes chaotic patterns as they proceed (as on the pelts of animals, but in test-tubes too), and it is a challenge to physical chemists to account for the patterns by establishing their complex mechanisms and computing their complex and sometimes strikingly beautiful outcomes.

Chapter 7
Investigating matter

Physical chemistry lies at the heart of one of chemistry's principal applications and achievements: the identification of the substances present in a sample and the determination of their abundances and structures. As these techniques increase in sophistication, physical chemists are ever more deeply involved in the interpretation of the rich data that they provide. They make use of many of the concepts that I have introduced in the preceding chapters, contributing through quantum theory (in spectroscopy), thermodynamics (through thermochemistry and electrochemistry), and kinetics. In many cases, a small army of concepts is responsible for the technique and its interpretation.

Two items have contributed enormously and almost universally to the elaboration of classical techniques: the laser and the computer. The laser has refined spectroscopy and kinetics, and has played a role even in thermochemistry. Few pieces of apparatus beyond the simplest are not touched either for control or interpretation by a computer, and the computer itself has enabled the entire field of computational chemistry.

Spectroscopy

Light, and electromagnetic radiation in general, brings information from the innermost parts of atoms and molecules.

The detection and analysis of the wavelengths and frequencies present in that light has been the source of an enormous amount of information and the techniques based on it continue to be developed. There are four flavours of spectroscopy—emission, absorption, scattering, and resonance—and different techniques for the different regions of the electromagnetic spectrum.

In *emission spectroscopy*, which is largely confined to the identification and study of atoms, light is emitted from atoms that have been energetically excited and then emit a photon as they collapse back to states of lower energy. The yellow light emitted by sodium atoms in street lighting is an example. As well as identifying the presence of elements in a sample, this kind of atomic spectroscopy gives very detailed information about the internal structures of atoms and was the original stimulus of the application of quantum mechanics to atoms.

Absorption spectroscopy is mainstream spectroscopy. In it, a beam of radiation of variable but precisely known frequency is passed through a sample and the attenuation of its intensity is recorded. There are three main sub-varieties. *Microwave spectroscopy* monitors the absorption of microwave radiation (radiation of wavelengths of several centimetres) by molecules that are free to rotate in the gas phase. It relies on the excitation of rotations and provides very precise information about bond lengths and bond angles. *Infrared spectroscopy* makes use of the absorption of infrared radiation by the vibrations of molecules and is used to identify the molecule from its vibrational fingerprint. *Ultraviolet/ visible spectroscopy* ('UV/vis spectroscopy') similarly monitors absorption, but now towards shorter wavelenths and higher frequencies. Absorption at these frequencies is due to the excitation of electrons out of the orbitals they occupy in the ground state. The sudden redistribution of electron density stimulates the molecule to burst into vibration and that bursting into vibration can also cause the molecule to rotate at different rates. The consequences of the stimulation of vibration and rotation when an electron is excited in a molecule can be detected in the gas phase and used to extract information about the strengths and stiffness of bonds; in solution they blur the absorption and give rise to broad bands in the spectrum.

I referred to scattering as one of the flavours. This is the domain of *Raman spectroscopy*, which was invented by Chandrasekhara Raman (1888–1970) and Kariamanickam Krishnan (1898–1961) in 1928 and for which Raman but not Krishnan quickly received a Nobel Prize (for physics) in 1930. The 'Raman effect' is the inelastic scattering of light by molecules. In the technique, photons of light (and ultraviolet radiation) are directed on to the sample and are scattered by its molecules. The scattered light has either a lower frequency if in its interaction with the molecules it loses energy to them or it has a higher frequency if it gains energy during the collision. Examination of the frequencies present in the scattered light reveals what energies the molecules can have and from that information can be deduced information about their identity and structure. Raman spectroscopy was little more than a laboratory backwater until the development of lasers because the intensity of the scattered radiation is low and concealed under the incident radiation unless the latter covers a very narrow band. Lasers provide intense, almost single-frequency ('monochromatic') light, so both these problems are overcome and Raman spectroscopy has moved into the mainstream of spectroscopic techniques. Allied with microscopy, it is also a powerful technique for the detailed analysis of surfaces.

The fourth flavour, resonance, is so important that I give it its own section.

Magnetic resonance

We all use resonance every day. Resonance occurs when the natural frequency of a device is tuned to the same frequency as a stimulus, such as the electromagnetic radiation from a distant radio transmitter. The response of the device is greatest when it is so tuned, and all kinds of radio communication depend on it. In its simplest application in spectroscopy, the properties of molecules are modified by controlling their environment (in all the applications I shall describe, by changing the strength of a magnetic field) until they are brought into resonance with the frequency of an oscillating electromagnetic field.

The hugely important technique of *nuclear magnetic resonance* (NMR), which was invented by physicists in the 1940s but quickly adopted by chemists to the point that no self-respecting laboratory can be without it, is a radiofrequency technique that uses (in modern applications, superconducting) magnets to influence the molecules it investigates. Many people have been the sample in an NMR observation, for it is used as a diagnostic technique as *magnetic resonance imaging* (MRI), where the 'nuclear' has been deleted from the name so as not to frighten the squeamish.

The 'nuclear' in NMR has nothing to do with nuclear radiation, only with nuclei. I mentioned in Chapter 1 that electrons have the property evocatively named 'spin'. Many nuclei have the same property: a hydrogen nucleus, the proton, has spin; a carbon nucleus (specifically the common isotope carbon-12) does not. A charged spinning body behaves like a bar magnet, so a nucleus with spin does so too. The north–south direction of the magnet depends on the direction of its spin, which in the case of a proton (like an electron) can be either clockwise or counterclockwise. It follows that protons (and nuclei with similar spin) will have different energies in an applied magnetic field depending on whether the bar magnet is north-up or north-down, and the energy separation will depend on the strength of the applied field (Figure 24). It is the ability to vary that energy separation that constitutes the 'tuning' in NMR. For the typical magnetic fields used in modern NMR, electromagnetic radiation in the region of 500 MHz is common, which lies in the 'radiofrequency' region of the spectrum (a little above FM radio which is close to 100 MHz).

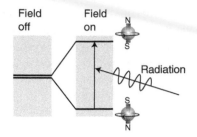

24. The basic process in NMR is to tune the separation of the two energy levels of a proton in a magnetic field to the frequency of a radiofrequency field. At resonance, the incident radiation is absorbed

The technique would be of little consequence in chemistry if all it achieved was flipping a spinning proton from one orientation to another. It acquires its extraordinary usefulness from two principal features (and in MRI a third). One feature is that protons in different parts of the same molecule experience slightly different magnetic fields from that applied: the applied field stirs up local currents in the molecules that augment or diminish the field, so resonances occur at different applied field strengths. These different 'chemical shifts' give valuable information about what groups of atoms are present. Second, a proton responds to the magnetic field generated by protons a bond or two away, and the interaction splits the single resonance absorption into characteristic patterns. This 'fine structure' helps to identify the molecule.

Modern NMR is far more sophisticated than this elementary account suggests, and physical chemists have contributed significantly to its development. Most NMR spectrometers use pulses of radiofrequency radiation to twist the spins of groups of protons into new orientations and to observe how protons twisted into high-energy orientations fall back into lower energy orientations. These are the so-called *Fourier-transform NMR* (FT-NMR) spectrometers, which use mathematical procedures to extract spectra from the data obtained from the sequence of pulses.

Clever pulse sequences, the replacement of non-magnetic carbon-12 nuclei by magnetic carbon-13 nuclei, extending the range of spectrometers to observe the resonance of phosphorus and fluorine nuclei, and inserting magnetic ions into the structure in known locations are all used not only to identify complex molecules but also to establish their structures. This application augments the information from X-ray diffraction (Chapter 4) because structures can be determined in natural aqueous environments typical of the interiors of biological cells, which might differ significantly from the structures adopted when the molecules lie artificially imprisoned together in crystals. Further developments in NMR technology are opening up the possibility of performing observations on samples of nanometre scales.

Nuclear magnetic resonance is also used to identify the motion of molecules in liquids and to establish the dynamical properties of the membranes that form the walls of biological cells. There are even hints that NMR might form one technique for realizing a quantum computer; so there might come a day when an NMR spectrometer understands what it is doing. The version of NMR used in MRI scans enables physicians to examine soft tissue non-invasively: it depends on exposing organisms to carefully controlled magnetic and radiofrequency electromagnetic fields, detecting the relaxation times of protons in different environments, and reconstructing three-dimensional images of the distribution of protons.

I called this section 'magnetic resonance' not 'nuclear magnetic resonance' because there are other magnetic resonance spectroscopies. Thus, *electron paramagnetic resonance* (EPR; or *electron spin resonance*, ESR) is similar to NMR but makes use of the magnetic properties of electrons. It is confined to molecules that have an unpaired electron (these include the radicals I mentioned in Chapter 6), so is far less widely applicable than NMR. However, it gives valuable information on these species and on some types of biological molecules (haemoglobin, for instance).

Mass spectrometries

A 'mass spectrum' is a totally different animal from the spectra I have mentioned so far, for the spectrometer used to obtain it detects fragments of molecules that are distinguished by their different masses. The principle involved is straightforward: smash a molecule into pieces, accelerate the ions to a speed that depends on their mass, and use a device that detects these different masses by deflecting them with an electric or magnetic field (the deflection also depends on the mass and speed) and recording the fragments that make the journey to a detector. The task before the experimenter is then to infer the identity and structure of the original molecule from the collection of fragments.

Mass spectrometry is widely used in organic chemistry to help identify compounds. One limitation that restricted its application to the large molecules of such importance in biology is the difficulty of getting enough of them present as a vapour so that the fragmentation of them by the impact of an electron beam was viable. That difficulty has largely been solved by embedding them in a polymeric material and then blasting it and the molecules into space by vaporizing it with a laser beam.

I called this section 'spectrometries' in the plural because I wanted to sneak in a mention of a variation of mass spectrometry that is more central to physical chemistry than is conventional mass spectrometry. *Photoelectron spectroscopy* (PES) is used both to explore the energies with which electrons are bound inside molecules and to identify species on surfaces.

In the technique, molecules are exposed to ultraviolet radiation that causes the ejection of electrons; these electrons replace the molecular fragments of conventional mass spectrometry, and otherwise the procedure is the same. The ejected electrons are accelerated and focused on a detector. By varying the strength of the field used to deflect them, the energy with which they left the molecule can be inferred. That energy is the difference between the energy of the photon of radiation that ejected them and the strength with which they were bound, so the latter can be deduced. This type of observation augments the calculations of computational chemistry and acts as an experimental verification of the energies that are computed as well as building up a more complete picture of the electronic structure of the molecule than ultraviolet spectroscopy can provide alone.

In a modification with the pedestrian name *electron spectroscopy for chemical analysis* (ESCA) the ultraviolet radiation of conventional PES is replaced by more energetic X-rays. Photons of X-rays are so energetic that they can eject the electrons that lie very close to nuclei. These electrons don't take part very much in chemical bonding so they have almost the same energy as in the unbound atom, and therefore the detected energies are characteristic of the element regardless of its state of combination and therefore can be used to identify the matter present.

Surface studies

Surfaces, despite being the outward show of a solid, have proved very difficult to study yet are of great importance, especially in catalysis. As I remarked in Chapter 6, catalytic action, the acceleration of chemical reactions, takes place on them. As I also mentioned there, their study was transformed some years ago by the introduction of a new technique and its elaborations.

The transformative technique is *scanning tunnelling microscopy* (STM). The key word here is 'tunnelling', which is a term borrowed from quantum mechanics and denotes going where classical mechanics forbids. The technique, as perhaps is true of most techniques that rely so centrally on quantum mechanics, is really most extraordinary and perhaps unlikely. In it, a metal probe is pulled out to form a very fine point, which is then dragged in a series of straight lines and at

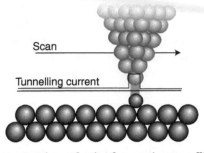

25. The process occurring at the tip of a scanning tunnelling microscope: the flow of current across the gap is highly sensitive to the width of the gap, and even atom-scale variation can be detected

constant height across the surface of the sample (there are variations in this procedure, but I will focus on this one); this is the 'scanning' part of the name. At the same time, the electric current passing through the tip is monitored and its variation used to map the rise and fall of the surface (Figure 25). The key point is that it is classically forbidden for electrons to jump across the gap between the tip of the probe and the surface, but quantum mechanics allow the electrons to cross the gap (this is the 'tunnelling' part of the name). This tunnelling is extremely sensitive to the width of the gap; it is so sensitive that it is sensitive to the rise and fall of the surface on an atomic scale, and the portrayal of the surface that results shows the individual atoms of the surface lying in serried ranks, with chasms and cliffs atoms deep and high, and reveals the shapes of individual molecules attached to the surface (this is the 'microscopy' of the name). Atoms, through STM, have at last become visible (Figure 26).

A variation of STM turns the surface chemist from passive observer into active participant. In *atomic force microscopy* (AFM) the tiny probe is used to nudge atoms around on the surface and in effect to build individual molecules atom by atom.

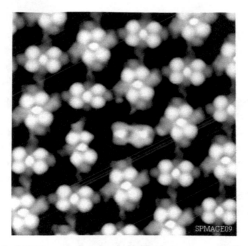

26. A typical STM scan, showing individual atoms on a surface

Surface chemists, the physical chemists who study surfaces, still continue to use classical techniques to study surfaces because the arrangement of atoms is not the only kind of information they need. A crucial part of the understanding that a surface plays in the promotion of chemical reactions is how much reactant adsorbs on the surface and its nature when it is there. Physical chemists have built models of the processes taking place on surfaces, have found relations between the amount adsorbed and the pressure of the overlying material in the gas, and have studied the energetics of adsorption. They find it initially helpful to distinguish between 'physisorption', when the molecule attaches unchanged to the surface, and 'chemisorption', when the adsorption occurs with the breaking and formation of covalent bonds and in some cases is accompanied by the fragmentation of the molecule. Chemisorption is the key to catalysis, for a molecule at least partially torn apart is ripe for reaction.

The surface area exposed by a catalyst is important for its function, for it is normally the case that the greater that area, the more effective is the catalyst. Surface chemists have methods for determining the area and of making it greater. One procedure is to synthesize so-called 'microporous' materials which (as mentioned in Chapter 4) are solids so riddled with pores, chambers, and channels that they are effectively all surface, with a gram of material having the surface area the size of a tennis court. Graphene, the material that has caused so much excitement on account of its unique mechanical, optical, and electrical properties and which has already won a Nobel Prize (for the physicists Andre Geim and Konstantin Novosolev in 2010) is essentially a single layer of chickenwire-like hexagonally arranged carbon atoms, as in the solid graphite. It is almost pure surface, and currently the target of the exploration of much surface chemistry.

Lasers

Lasers have transformed many aspects of physical chemistry and are common features in laboratories. Three of the characteristics of the radiation they produce are responsible, and I have touched on them in Chapter 6.

One is their intensity. Each pulse of radiation can be regarded as a hurtling swarm of an enormous number of photons. This feature rejuvenated Raman spectroscopy for even a small proportion of inelastically scattered photons from that swarm can still give rise to a reasonably intense beam to detect, record, and analyse. Intensity also brings into range so-called 'non-linear optical phenomena', in which more than one photon might interact simultaneously with a molecule instead of just the normal one-photon/one-molecule interaction, and give rise to new phenomena and new ways to explore its structure. The new phenomena include the separation of isotopes by laser irradiation and new opportunities for using laser radiation in chemical synthesis. Even the light that is scattered from molecules without change of frequency, the so-called 'Rayleigh scattering', can be employed to assess the sizes of the large molecules typical of polymers and to examine the motion of molecules in fluids.

Physical chemists also make use of the monochromaticity of laser radiation, the fact that it consists of a very narrow range of wavelengths and frequencies. Not only does that narrowness make Raman spectroscopy more feasible than when it had to rely on conventional optical sources but it also gives the experimenter great precision in selecting the state to which molecules are excited in a photochemical experiment.

I also explained in Chapter 6 how the sharpness of the pulses from lasers can be employed to investigate processes that occur very rapidly. A process can be stimulated with a short, sharp pulse, and then examined almost instantly later by sending in another pulse and monitoring its absorption. Thus investigations of processes taking place in fractions of a second (even attoseconds, 1 as = 10^{-18} s) can be made and extraordinarily detailed pictures of chemical reactions constructed.

The intense field of a laser beam has another, entirely different application: when focused down to almost a point, it can act as 'optical tweezers' and hold tiny particles of matter in place for them to be studied individually. Other spectroscopic techniques are also emerging which enable single molecules to be studied. This type of study is especially important for watching the possibly complex behaviour of biological molecules as they participate in biochemical processes.

Computers

Computers have profoundly affected the instrumentation in modern physical chemistry laboratories and have applications well outside their intrinsic contribution to computational chemistry, the quantum mechanical calculation of molecular structure, including the still vexed question of how the long amino acid chains of protein molecules fold into the precise, semi-rigid shapes that are crucial to their function. The latter falls, to some extent, into the field of 'molecular mechanics' where Newton's laws of classical physics are used (the quantum mechanics being plain too difficult) to predict how a molecule or parts of a big molecule move under the influence of the forces acting within the molecule. Physical chemists have identified most sources of these forces. The difficulty with coping with them is that some are long-range, acting over the width of a molecule, some are short-range, acting only when parts of molecules come into contact, and some aren't really forces at all. By that last enigmatic remark I mean that any water around parts of a biological molecule can influence its motion as though it were exerting a force. Building all these effects into the model is very demanding and remains a focus of much current research.

Computers control spectrometers and diffractometers (for X-ray diffraction). Modern spectroscopic techniques commonly use indirect procedures for obtaining spectra, the data needing to be manipulated by quite extensive computation. The useful analogy that illustrates these 'Fourier transform' techniques is that of determining the frequencies (the notes) present in a piano. One way is to play each note in turn and to record its presence. An alternative is to drop the piano and record the awful noise as it, the noise, fades away. The individual notes that contribute to that noise can be identified by taking the recording of the noise and performing on it the same kind of mathematical transformation, the Fourier analysis. Such procedures, which are used in several branches of spectroscopy, including NMR, greatly accelerate the collection of the spectral data and provide highly detailed information.

The current challenge

Old techniques become more sophisticated and reveal information that was out of the range of their predecessors; new techniques open our intellectual eyes to new properties. Both bring in their train problems of interpretation and the elaboration of old models of Nature. Physical chemistry is a happy marriage of experiment and theory, the latter being inspired by the former and the theory inspiring new experiments. Often those experiments simply confirm what has been supposed or add quantitative spine to qualitative ideas. But often they open up whole new landscapes for physical chemistry to enter and explore.

The laser goes on producing surprises and opportunities, such as exploring ever shorter timescales and enabling physical chemists to examine single molecules to understand both their structures and their dynamical behaviour. Computing, especially high-performance computing, where vast computational resources can be brought to bear to generate simulations of elaborate systems, is undoubtedly the way forward and will bring opportunities for understanding that are beyond our, or at least my, imagining. Detailed information on an atomic scale is being obtained by using very intense radiation sources, such as the radiation generated by synchrotron sources (a national facility in many regions, as they are huge machines), including the extension of X-ray studies of proteins to their dynamical behaviour. Microscopy of various flavours is used in tandem with spectroscopies to extract detailed information about surface phenomena. Even new forms of matter, such as the special kind of matter, the so-called *Bose condensate*, that forms when miniscule collections of atoms are isolated and cooled to such low temperatures that quantum phenomena become dominant, are becoming open to investigation.

Not only do physical chemists contribute to the development of new techniques for the study of conventional, intriguing, novel, forms of matter, but they also contribute to the extraction of the information that these ever more sophisticated techniques provide. They are extending their domain of enquiry and elucidation to nanostructures, where entirely new kinds of instrumentation and theoretical analysis are needed, and increasingly to biology, where traditional phenomena collaborate, display complex behaviour, and cannot be understood without computer analysis and simulation. Physical chemists will never be left with nothing to do.

Appendix: the Periodic Table

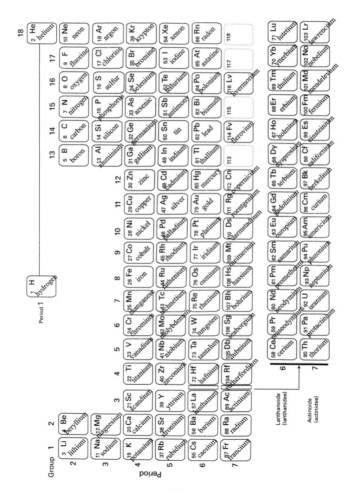

Further reading

I have elaborated many of the concepts discussed in this book in other books. In particular, I mention:

Four Laws that Drive the Universe (2007), reissued as *The Laws of Thermodynamics: A Very Short Introduction*, Oxford: Oxford University Press (2010).

What is Chemistry? (2013), reissued as *Chemistry: A Very Short Introduction*, Oxford: Oxford University Press (2015).

Reactions: The Private Life of Atoms, Oxford: Oxford University Press (2011).

Three of my textbooks provide deeper information, and in order of increasing sophistication are

Chemical Principles, with Loretta Jones and Leroy Laverman, New York: W. H. Freeman & Co. (2013).

Elements of Physical Chemistry, with Julio de Paula, Oxford and New York: Oxford University Press and W. H. Freeman & Co. (2013).

Physical Chemistry, with Julio de Paula, Oxford and New York: Oxford University Press and W. H. Freeman & Co. (2014).

Index

Expand your collection of
VERY SHORT INTRODUCTIONS

SOCIAL MEDIA
Very Short Introduction

Join our community
www.oup.com/vsi

- Join us online at the official Very Short Introductions **Facebook** page.
- Access the thoughts and musings of our authors with our online **blog**.
- Sign up for our monthly **e-newsletter** to receive information on all new titles publishing that month.
- Browse the full range of Very Short Introductions online.
- Read **extracts** from the Introductions for free.
- Visit our library of **Reading Guides**. These guides, written by our expert authors will help you to question again, why you think what you think.
- If you are a teacher or lecturer you can order inspection copies quickly and simply via our website.

ONLINE CATALOGUE
A Very Short Introduction

Our online catalogue is designed to make it easy to find your ideal Very Short Introduction. View the entire collection by subject area, watch author videos, read sample chapters, and download reading guides.

http://fds.oup.com/www.oup.co.uk/general/vsi/index.html